TECHNIQUES IN HIGH PRESSURE NEUTRON SCATTERING

TECHNIQUES IN HIGH PRESSURE NEUTRON SCATTERING

Editor
Khamisi Zaman

scitus
academics

Techniques in High Pressure Neutron Scattering
Edited by **Khamisi Zaman**

ISBN: 978-1-68117-717-5
Library of Congress Control Number: 2016958823

SCITUS Academics LLC,
616, Corporate Way, Suite 2, 4766,
Valley Cottage, NY 10989

www.scitusacademics.com

Notice

Printed in United States of America on Acid Free Paper

PREFACE

Neutron scattering experiments determine the atomic and/or magnetic structure of a material. Neutron scattering as an experimental technique is used in crystallography, physics, physical chemistry, biophysics, and materials research. It is practiced at research reactors and spallation neutron sources that provide neutron radiation of sufficient intensity. Neutron diffraction (elastic scattering) is used for determining structures; inelastic neutron scattering is used for the study of atomic vibrations and other excitations. It is the ideal method to study a wide range of important and common materials like ice and ammonia. Knowledge of the positions of the atoms in systems like these when they are compressed helps our understanding of topics as diverse as the behavior of proteins and the origins of the magnetic fields of planets like Uranus and Neptune. Until 1990, neutron diffraction was limited to pressures below 30,000 atmospheres. This was a very low pressure when compared to other structural techniques like x-ray diffraction and light-scattering which at that time were achieving pressures in excess of 1 million atmospheres. The ability to manipulate structure and properties using pressure has been well known for many centuries. Diffraction provides the unique ability to observe these structural changes in fine detail on length-scales spanning atomic to nanometre dimensions. Amongst the broad suite of scattering tools available today, neutrons provide unique capabilities of fundamental importance. However, to date, the growth of neutron scattering under extremes of pressure has been limited by the weakness of available sources. In recent years, extensive investments have led to the construction of a new generation of neutron sources while existing facilities have been revitalized by upgrades. The timely convergence of these bright facilities with new pressure-cell technologies suggests that the field of high-pressure neutron science is on the cusp of substantial growth. It has increased the pressure range for neutron scattering by a factor of ten with the development of a novel large volume pressure cell.

Techniques in High Pressure Neutron Scattering presents contemporary methods and techniques that allow neutron scattering well beyond 10 GPa. It demonstrates how high pressure neutron scattering can be used to study the structure and dynamics of a wide range of materials. It will be of pertinent for newcomers and experienced high pressure scientists and engineers, researchers and practitioners, as well as advanced graduate students.

TABLE OF CONTENTS

1

CONCEPTS, INSTRUMENTATION AND TECHNIQUES OF NEUTRON ACTIVATION ANALYSIS

Lylia Hamidatou[1], Hocine Slamene[1], Tarik Akhal[1] and Boussaad Zouranen[1]

[1]Department of Neutron Activation Analysis, Nuclear Research Centre of Birine, Algeria

1. INTRODUCTION

Analytical science to develop the methodology for the investigation of properties and structure of matter at level of single nucleus, atom and molecule, and scientific analysis to determine either chemical composition or elemental contents in a sample are indispensable in basic research and development, as well as in industrial applications.

Following the discovery of neutron by J. Chadwick in 1932 (Nobel prize, 1935) and the results of F. Joliot and I. Curie in 1934, neutron activation analysis was first developed by G. Hevesy and H. Levi in 1936. They used a neutron source (226Ra + Be) and a radiation detector (ionization chamber) and promptly recognized that the element Dy (dysprosium) in the sample became highly radioactive after exposure to the neutron source. They showed that the nuclear reaction may be used to determine the elements present in unknown samples by measuring the induced radioactivity.

Thereafter, the development of the nuclear reactors in the 1940s, the application of radiochemical techniques using low resolution scintillation detectors like NaI (Tl) in the 1950s, the development of semiconductor detectors (Ge, Si, etc.) and multichannel analyzer in the 1960s, and the advent of computers and relevant software in the 1970s, the nuclear technique has advanced to become an important analytical tool for determination of many elements at trace level. In spite of the developments in other chemical techniques, the simplicity and selectivity, the speed of operation, the sensitivity and accuracy of NAA have become and maintained its role as a powerful analytical technique. In 2011, Peter Bode describes in his paper “Neutron activation analysis: A primary method of measurements”, the history of the development of NAA overall the world [1].

Nowadays, there are many elemental analysis methods that use chemical, physical and nuclear characteristics. However, a particular method may be favoured for a specific task, depending on the purpose. Neutron activation analysis (NAA) is very useful as sensitive analytical technique for performing both qualitative and quantitative multielemental analysis of major, minor and traces components in variety of terrestrial samples and extra-terrestrial materials. In addition, because of its accuracy and reliability, NAA is generally recognized as the "referee method" of choice when new procedures are being developed or when other methods yield results that do not agree. It is usually used as an important reference for other analysis methods. Worldwide application of NAA is so widespread it is estimated that approximately 100,000 samples undergo analysis each year.

The method is based on conversion of stable atomic nuclei into radioactive nuclei by irradiation with neutrons and subsequent detection of the radiation emitted by the radioactive nuclei and its identification. The basic essentials required to carry out an analysis of samples by NAA are a source of neutrons, instrumentation suitable for detecting gamma rays, and a detailed knowledge of the reactions that occur when neutrons interact with target nuclei. Brief descriptions of the NAA method, reactor neutron sources, and gamma-ray detection are given below.

This chapter describes in the first part the basic essentials of the neutron activation analysis such as the principles of the NAA method with reference to neutron induced reactions, neutron capture cross-sections, production and decay of radioactive isotopes, and nuclear decay and the detection of radiation. In the second part we illustrated the equipment requirements neutron sources followed by a brief description of Es-Salam research reactor, gamma-ray detectors, and multi-channel analysers. In addition, the preparation of samples for neutron irradiation, the instrumental neutron activation analysis techniques, calculations, and systematic errors are given below. Some schemes of irradiation facilities, equipment and materials are given as examples in this section.

Finally, a great attention will be directed towards the most recent applications of the INAA and k0-NAA techniques applied in our laboratory. Examples of such samples, within a selected group of disciplines are milk, milk formulae and salt (nutrition), human hair and medicinal seeds (biomedicine), cigarette tobacco (environmental and health related fields) and iron ores (exploration and mining).

All steps of work were performed using NAA facilities while starting with the preparation of samples in the laboratory. The activation of samples depends of neutron fluence rate in irradiation channels of the Algerian Es-Salam research reactor. The radioactivity induced is measured by gamma spectrometers consist of germanium based semiconductor detectors connected to a computer used as a multichannel analyser for spectra evaluation and calculation. Sustainable developments of advanced equipment, facilities and manpower have been implemented to establish a state of the art measurement capability, to implement several applications, etc.

2. NEUTRON ACTIVATION ANALYSIS

Neutron activation analysis (NAA) is a nuclear process used for determining the concentrations of elements in a vast amount of materials. NAA relies on excitation by neutrons so that the treated sample emits gamma-rays. It allows the precise identification and quantification of the elements, above all of the trace elements in the sample. NAA has applications in chemistry but also in other research fields, such as geology, archaeology, medicine, environmental monitoring and even in the forensic science.

2.1. Basis Principles

The sequence of events occurring during the most common type of nuclear reaction used for NAA, namely the neutron capture or (n, gamma) reaction, is illustrated in Figure 1. Creation of a compound nucleus forms in an excited state when a neutron interacts with the target nucleus via a non-elastic collision. The excitation energy of the compound nucleus is due to the binding energy of the neutron with the nucleus. The compound nucleus will almost instantaneously de-excite into a more stable configuration through emission of one or more characteristic prompt gamma rays. In many cases, this new configuration yields a radioactive nucleus which also de-excites (or decays) by emission of one or more characteristic delayed gamma rays, but at a much lower rate according to the unique half-life of the radioactive nucleus. Depending upon the particular radioactive species, half-lives can range from fractions of a second to several years.

In principle, therefore, with respect to the time of measurement, NAA falls into two categories: (1) prompt gamma-ray neutron activation analysis (PGNAA), where measurements take place during irradiation, or (2) delayed gamma-ray neutron activation analysis (DGNAA), where the measurements follow radioactive decay. The latter operational mode is more common; thus, when one mentions NAA it is generally assumed that measurement of the delayed gamma rays is intended. About 70% of the elements have properties suitable for measurement by NAA.

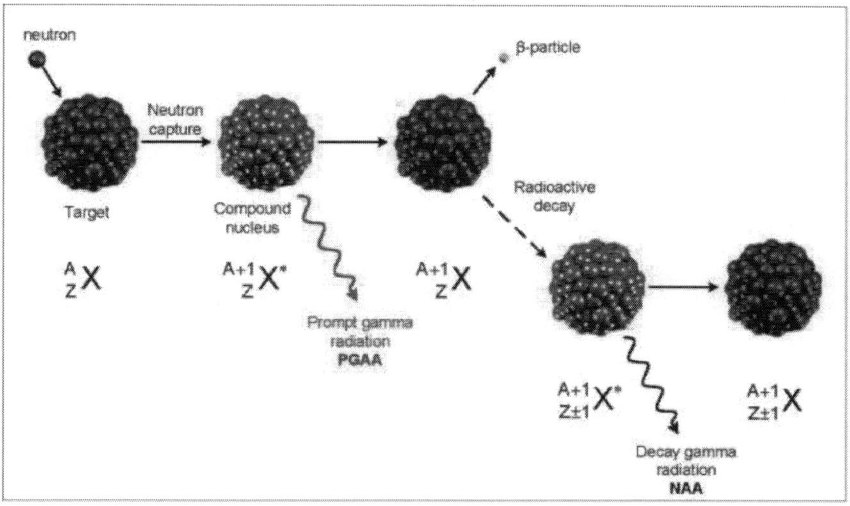

Figure 1. Diagram illustrating the process of neutron capture by a target nucleus followed by the emission of gamma rays.

The PGAA technique is generally performed by using a beam of neutrons extracted through a reactor beam port. Fluxes on samples irradiated in beams are in the order of one million times lower than on samples inside a reactor but detectors can be placed very close to the sample compensating for much of the loss in sensitivity due to flux. The PGAA technique is most applicable to elements with extremely high neutron capture cross-sections (B, Cd, Sm, and Gd); elements which decay too rapidly to be measured by DGAA; elements that produce only stable isotopes (e.g. light elements); or elements with weak decay gamma-ray intensities. 2D, 3D-analysis of (main) elements distribution in the samples can be performed by PGAA.

DGNAA (sometimes called conventional NAA) is useful for the vast majority of elements that produce radioactive nuclides. The technique is flexible with respect to time such that the sensitivity for a long-lived radionuclide that suffers from interference by a shorter-lived radionuclide can be improved by waiting for the short-lived radionuclide to decay or quite the contrary, the sensitivity for short-lived isotopes can be improved by reducing the time irradiation to minimize the interference of long-lived isotopes. This selectivity is a key advantage of DGNAA over other analytical methods.

In most cases, the radioactive isotopes decay and emit beta particles accompanied by gamma quanta of characteristic energies, and the radiation can be used both to identify and accurately quantify the elements of the sample. Subsequent to irradiation, the samples can be measured instrumentally by a high resolution semiconductor detector, or for better sensitivity, chemical separations can also be applied to reduce interferences. The qualitative characteristics are: the energy of the emitted gamma quanta (Eγ) and the half life of the nuclide ($T_{½}$). The quantitative characteristic is: the Iγ intensity, which is the number of gamma quanta of energy Eγ measured per unit time.

The n-gamma reaction is the fundamental reaction for neutron activation analysis. For example, consider the following reaction:

58Fe +1n →59Fe + Beta-+ gamma rays58Fe +1n →59Fe + Beta-
+ gamma rays

^{58}Fe is a stable isotope of iron while ^{59}Fe is a radioactive isotope. The gamma rays emitted during the decay of the ^{59}Fe nucleus have energies of 142.4, 1099.2, and 1291.6 KeV, and these gamma ray energies are characteristic for this nuclide (see figure 2) [2]. The probability of a neutron interacting with a nucleus is a function of the neutron energy. This probability is referred to as the capture cross-section, and each nuclide has its own neutron energy-capture cross-section relationship. For many nuclides, the capture cross-section is greatest for low energy neutrons (referred to as thermal neutrons). Some nuclides have greater capture cross-sections for higher energy neutrons (epithermal neutrons). For routine neutron activation analysis we are generally looking at nuclides that are activated by thermal neutrons.

The most common reaction occurring in NAA is the (n,γ) reaction, but also reactions such as (n,p), (n,α), (n,n′) and (n,2n) are important. The neutron cross section, σ, is a measure for the probability that a reaction will take place, and can be strongly different for different reaction types, elements and energy distributions of the bombarding neutrons. Some nuclei, like 235U are fissionable by neutron capture and the reaction is denoted as (n,f), yielding fission products and fast (highly energetic) neutrons [1].

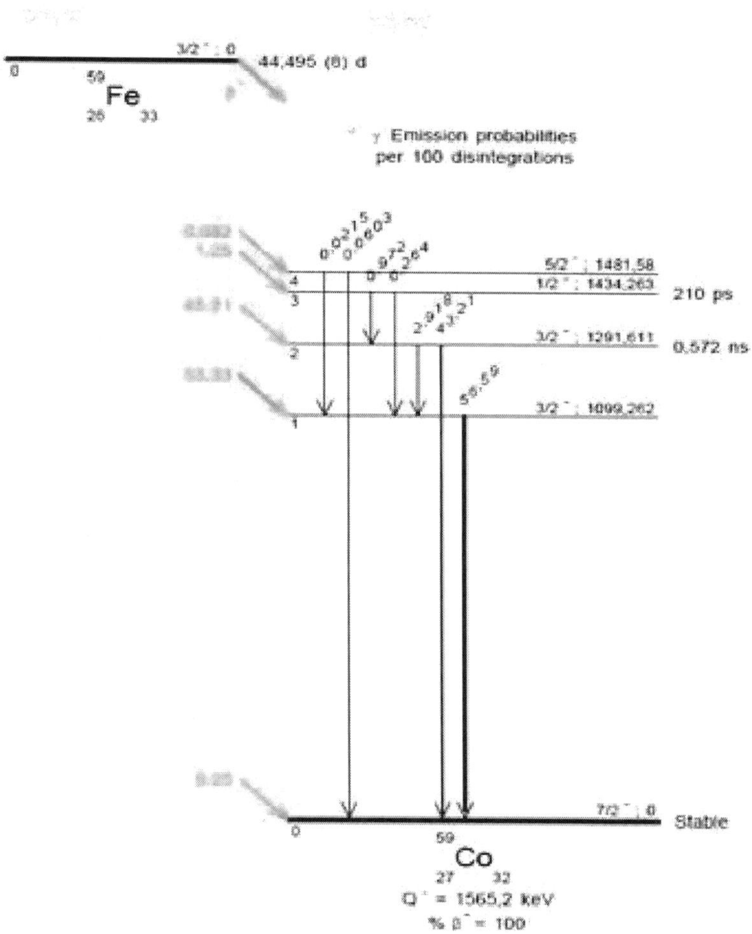

Figure 2. Decay scheme of ^{59}Fe.

Neutrons are produced via

- Isotopic neutron sources, like ^{226}Ra(Be), ^{124}Sb(Be), ^{241}Am(Be), ^{252}Cf. The neutrons have different energy distributions with a maximum in the order of 3–4 MeV; the total output is typically 10^5–10^7 s^{-1} GBq^{-1} or, for ^{252}Cf, 2.2 10^{12} $s^{-1}g^{-1}$.

- Particle accelerators or neutron generators. The most common types are based on the acceleration of deuterium ions towards a target containing either deuterium or tritium, resulting in the reactions 2H(2H,n)3He and 3H(2H,n)4He, respectively. The first reaction, often denoted as (D,D), yields monoenergetic neutrons of

2.5 MeV and typical outputs in the order of 10^8–10^{10} s^{-1}; the second reaction (D,T) results in monoenergetic neutrons of 14.7 MeV and outputs of 10^9–10^{11} s^{-1}.

- Nuclear research reactors. The neutron energy distribution depends on design of the reactor and its irradiation facilities. An example of an energy distribution in a light water moderated reactor is given in Fig. 2.3 from which it can be seen that the major part of the neutrons has a much lower energy distribution that in isotopic sources and neutron generators. The neutron output of research reactors is often quoted as neutron fluence rate in an irradiation facility and varies, depending on reactor design and reactor power, between 10^{15} and 10^{18} m^{-2} s^{-1}.

Owing to the high neutron flux, experimental nuclear reactors operating in the maximum thermal power region of 100 kW -10 MW with a maximum thermal neutron flux of 10^{12}-10^{14} neutrons $cm^{-2}s^{-1}$ are the most efficient neutron sources for high sensitivity activation analysis induced by epithermal and thermal neutrons. The reason for the high sensitivity is that the cross section of neutron activation is high in the thermal region for the majority of the elements. There is a wide distribution of neutron energy in a reactor and, therefore, interfering reactions must be considered. In order to take these reactions into account, the neutron spectrum in the channels of irradiation should be known exactly. E.g. if thermal neutron irradiations are required, the most thermalized channels should be chosen.

Although there are several types of neutron sources (reactors, accelerators, and radioisotopic neutron emitters) one can use for NAA, nuclear reactors with their high fluxes of neutrons from uranium fission offer the highest available sensitivities for most elements. Different types of reactors and different positions within a reactor can vary considerably with regard to their neutron energy distributions and fluxes due to the materials used to moderate (or reduce the energies of) the primary fission neutrons. This is further elaborated in the title "Derivation of the measurement equation". In our case, the NAA method is based on the use of neutron flux in several irradiation channels of Es-Salam Research reactor. In 2011, Hamidatou L et Al., reported "Experimental and MCNP calculations of neutron flux parameters in irradiation channel at Es-Salam reactor" the core modelling to calculate neutron spectra using experimental and MCNP approaches. The Es-Salam reactor was designed for a thermal power output of 15 Mw, with 72 cylindrical cluster fuel elements; each fuel element consists of 12 cylindrical rods of low enriched UO2. In addition the both of fuel throttle tube of the cluster and fuel element tube encloses heavy water as moderator and coolant. The fuel elements are arranged on a heavy water square lattice. The core of the reactor is constituted by a grid containing 72 fuel elements, 12 rods for reactivity control and two experimental channels.

There is also a heavy water in the middle of the core including five experimental channels called inner reflector, In addition, all fuel elements have a reflector at each end called upper and lower reflector. The core is reflected laterally by heavy water maintained in aluminium tank followed by the graphite.

2.2. Neutron Activation Analysis Procedure

In the majority of INAA procedures thermal reactor neutrons are used for the activation: neutrons in thermal equilibrium with their environment. Sometimes activation with epithermal reactor neutrons (neutrons in the process of slowing down after their formation from fission of 235U) is preferred to enhance the activation of elements with a high ratio of resonance neutron cross section over thermal neutron cross section relatively to the activation of elements with a lower such a ratio. In principle materials can be activated in any physical state, viz. solid, liquid or gaseous. There is no fundamental necessity to convert solid material into a solution prior to activation; INAA is essentially considered to be a non-destructive method although under certain conditions some material damage may occur due to thermal heating, radiolysis and radiation tracks by e.g. fission fragments and α-radiation emitting nuclei. It is essential to have more than two or three qualified full-time member of the staff with responsibility for the NAA facilities. They should be able to control the counting equipment and have good knowledge of basic principles of the technique. In addition, the facility users and the operators must establish a good channel of communication. Other support staff will be required to maintain and improve the equipment and facility. It seems, therefore, a multi-disciplinary team could run the NAA system well.

The analytical procedure is based on four steps:

Step 1: sample preparation (Figure 3) means in most cases only heating or freeze drying, crushing or pulverization, fractionating or pelletizing, evaporation or pre-concentration, put through a sieve, homogenising, weighing, washing, check of impurities (blank test), encapsulation and sealing irradiation vial, as well as the selection of the best analytical process and the preparation of the standards. The laboratory ambiance is also important for preservation and storage of the samples. Standardization is the basis for good accuracy of analytical tools and often depends on particular technology, facility and personnel. For production of accurate data, careful attention to all possible errors in preparing single or multi-element standards is important, and standards must be well chosen depending on the nature of the samples.

Figure 3. Some instruments and materials used for the sample preparation.

Step 2: irradiation of samples can be taken from the various types of neutron sources according to need and availability. For the INAA, one pneumatic transfer system installed in the horizontal channel at Es-Salam research reactor for short irradiation of samples (Figure 4). In addition, two vertical channels located in different sites of the heavy water moderator and the graphite reflector have been used for long irradiations. The neutron spectrum parameters at different irradiation channels such as alpha, f, Tn, etc are experimentally determined using cadmium ratio, cadmium cover, bare triple monitor and bi-isotopic methods using HΦgdhal convention and Westcott formalism Table 1 and Table 2. The calibration of the irradiation positions has been carried out to implement the k_0-NAA in our laboratory.

Table 1. The parameters α, f and $r(\alpha)\sqrt{T_n / T_0}$ obtained by different methods.

method	α	f	$r(\alpha)\sqrt{T_n / T_0}$
Cd-ratio	0.026±0.012	28.4±1.6	0.038±0.004
Cd-covered	0.024±0.010	28.7±2.1	-
Bare triple monitor	0.030±0.008	28.6±1.8	-
Bare bi-isotopic	-	29.5±2.5	0.036±0.003
Average	0.027±0.010	28.8±2.0	0.037±0.003

Table 2. Neutron spectrum parameters in the irradiation site at es-Salam research reactor.

parameter	α	f	Tn(°C)	Rcd(Au)	$r(\alpha)\sqrt{T_n / T_0}$
Measured value	0.027±0.010	28.8±2.0	34±1.8	2.93±0.32	0.037±0.003

Step 3: after the irradiation the measurement is performed after a suitable cooling time (t_c). In NAA, nearly exclusively the (energy of the) gamma radiation is measured because of its higher penetrating power of this type of radiation, and the selectivity that can be obtained from distinct energies of the photons - differently from beta radiation which is a continuous energy distribution. The interaction of gamma- and X-radiation with matter results, among others, in ionization processes and subsequent generation of electrical signals (currents) that can be detected and recorded.

The instrumentation used to measure gamma rays from radioactive samples generally consists of a semiconductor detector, associated electronics, and a computer-based multi-channel analyzer (MCA/computer).

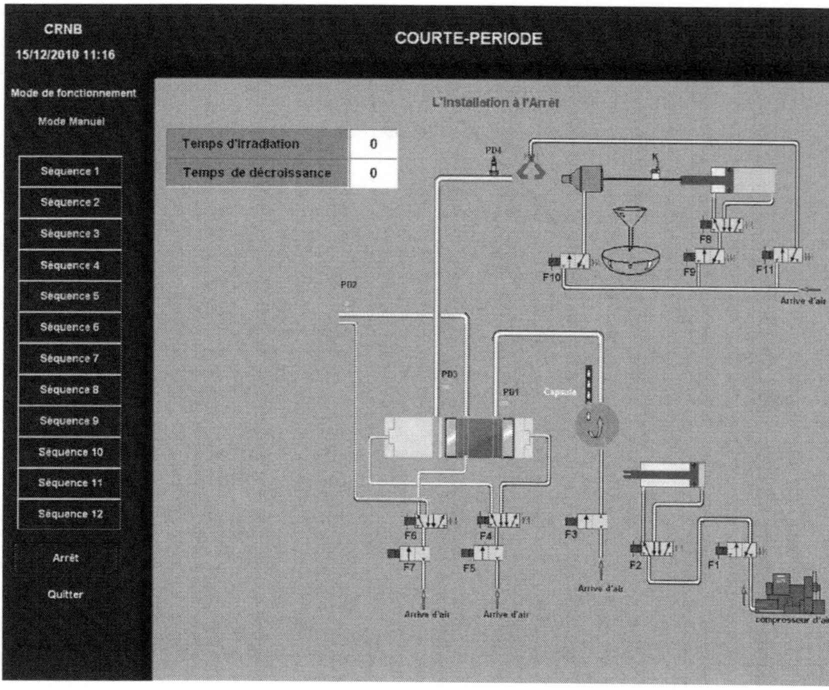

Figure 4. Pneumatic system for short irradiations using a thermal neutron flux at Es-Salam research reactor.

Most NAA labs operate one or more hyper-pure germanium (HPGe) detectors, which operate at liquid nitrogen temperature (77 K). Although HPGe detectors come in many different shapes and sizes, the most common shape is coaxial. These detectors are very useful for measurement of gamma rays with energies in the range from about 60 keV to 3.0 MeV. The two most important characteristics a HPGe detector are its resolution and efficiency. Other characteristics to consider are peak shape, peak-to-Compton ratio, pulse rise time, crystal dimensions or shape, and price. The detector's resolution is a measure of its ability to separate closely spaced peaks in the spectrum, and, in general, the resolution is specified in terms of the full width at half maximum (FWHM) of the 122 keV photopeak of ^{57}Co and the 1,332 keV photopeak of ^{60}Co. For most NAA applications, a detector with 0.5 keV resolution or less at 122 keV and 1.8 keV or less at 1,332 keV is sufficient. Detector efficiency for a given detector depends on gamma-ray energy and the sample and detector geometry, i.e. subtended solid angle. Of course, a larger volume detector will have a higher efficiency.

At Es-Salam NAA Lab, four gamma-ray spectrometers of Canberra for which one of them consists of a HPGe detector 35% relative efficiency connected with Genie 2k Inspector and the three other spectrometers are composed of detectors (30, 35 and 45 % relative efficiency) connected with a three Lynx® Digital Signal Analyser, It is a 32K channel integrated signal analyzer based on advanced digital signal processing (DSP) techniques. All spectrometers operate with Genie™2000 spectroscopy software. A radiation detector therefore consists of an absorbing material in which at least part of the radiation energy is converted into detectable products, and a system for the detection of these products.Figure 5 illustrates Gamma-ray spectroscopy systems. The detectors are kept at liquid nitrogen temperatures (dewers under cave). The boxes in the left and in the right of the computer are the Lynx Digital Spectrometer Processing.

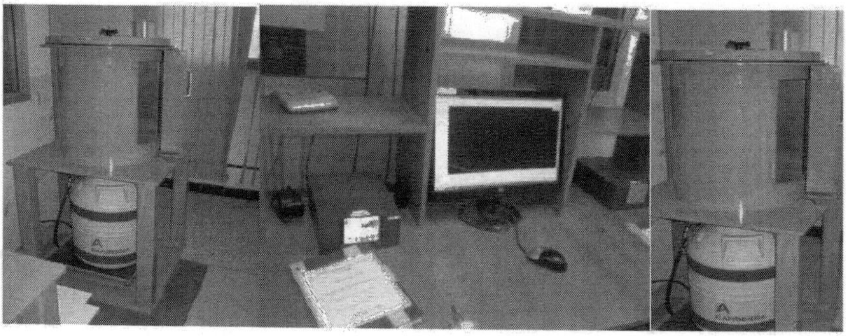

Figure 5. Gamma-ray spectroscopy systems in NAA/CRNB laboratory.

Step 5: Measurement, evaluation and calculation involve taking the gamma spectra and the calculating trace element concentrations of the sample and preparation of the NAA report.

In this part of work, Peter bode describes clearly in his paper [1] the analysis procedure of gamma-spectrum to the determination of the amount of element in sample. The acquisition of gamma spectrumFig.6 and Fig.7 via the spectroscopy system Fig. 5 is analyzed to identify the radionuclides produced and their amounts of radioactivity in order to derive the target elements from which they have been produced and their masses in the activated sample. The spectrum analysis starts with the determination of the location of the (centroids of the) peaks. Secondly, the peaks are fitted to obtain their precise positions and net peak areas. The Analytical protocol adopted in our NAA laboratory is presented inFig.8.

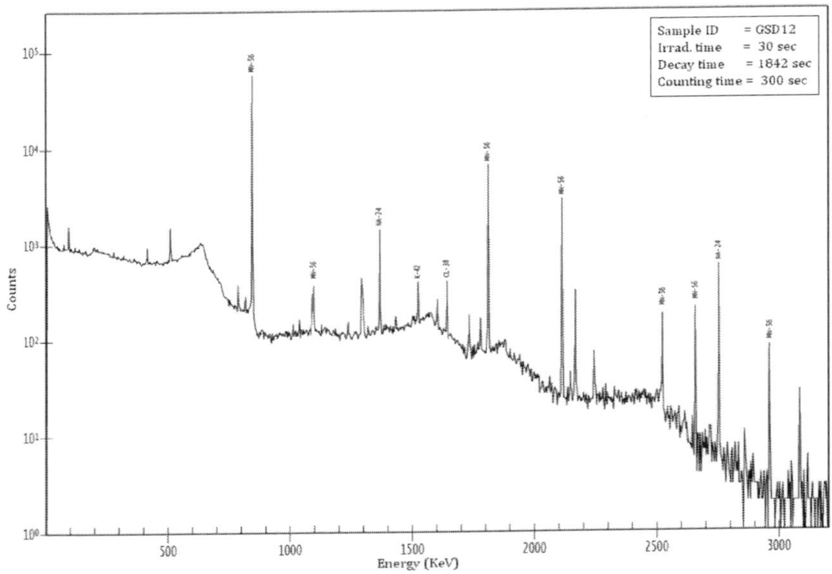

Figure 6. Gamma-ray spectrum showing several short-lived elements measured in a CRM-DSD-12 standard irradiated at Es-salam research reactor for 30 seconds, decayed for 30.7 minutes, and counted for 5 minutes with an HPGe detector.

The positions – often expressed as channel numbers of the memory of a multi-channel pulse height analyzer – can be converted into the energies of the radiation emitted; this is the basis for the identification of the radioactive nuclei. On basis of knowledge of possible nuclear reactions upon neutron activation, the (stable) element composition is derived. The values of the net peak areas can be used to calculate the amounts of radioactivity of the radionuclides using the full energy photopeak efficiency of the detector.

The amounts (mass) of the elements may then be determined if the neutron fluence rate and cross sections are known. In the practice, however, the masses of the elements are determined from the net peak areas by comparison with the induced radioactivity of the same neutron activation produced radionuclides from known amounts of the element of interest. The combination of energy of emitted radiation, relative intensities if photons of different energies are emitted and the half life of the radionuclide is unique for each radionuclide, and forms the basis of the qualitative information in NAA. The amount of the radiation is directly proportional to the number of radioactive nuclei produced (and decaying), and thus with the number of nuclei of the stable isotope that underwent the nuclear reaction. It provides the quantitative information in NAA.

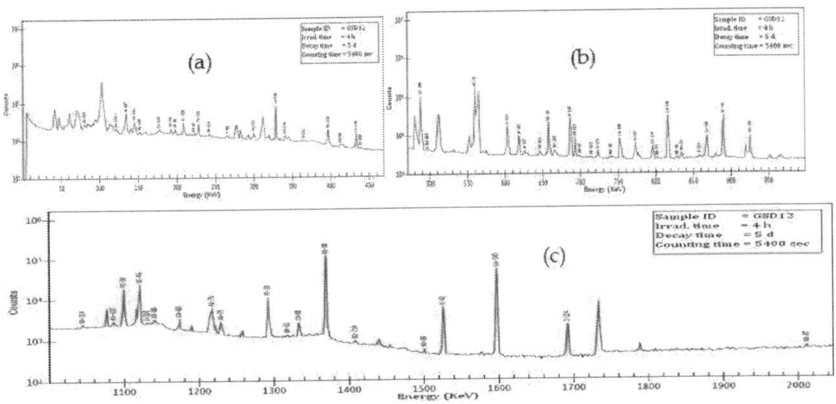

Figure 7. Gamma-ray spectrum (a) from 0 to 450 keV, (b) from 450 to 1000 keV and (c) from 1000 to 2000 keV: showing medium- and long-lived elements measured in a sample of CRM-GSD-12 standard irradiated at Es-salam research reactor for 4 hours, decayed for 5 days, and counted for 90 minutes on a HPGe detector.

The measured in NAA – the quantity intended to be measured – is the total mass of a given element in a test portion of a sample of a given matrix in all physico-chemical states. The quantity 'subject to measurement' is the number of disintegrating nuclei of a radionuclide. The measurement results in the number of counts in a given period of time, from which the disintegration rate and the number of disintegrating nuclei is calculated; the latter number is directly proportional to the number of nuclei of the stable isotope subject to the nuclear reaction, and thus to the number of nuclei of the element, which finally provides information on the mass and amount of substance of that element (see Eq. 16). An example of typical ranges of experimental conditions is given in Table 3 [1].

In practice, our laboratory proceeds in the treatment of spectra and calculation of elemental concentrations of analyzed samples according the approach illustrated in figure 8.

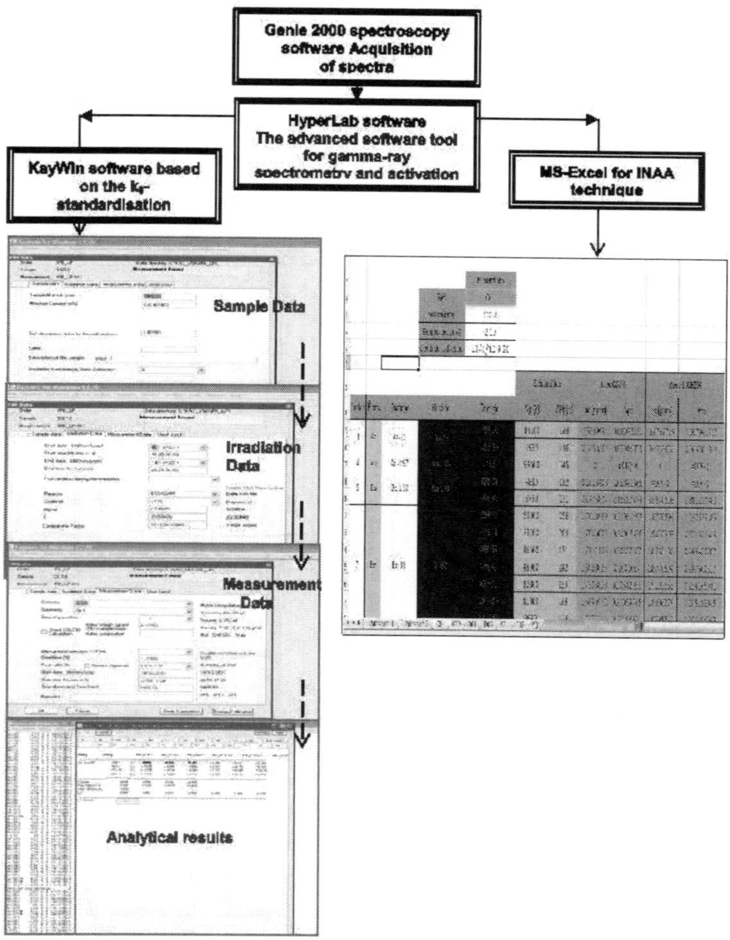

Figure 8. Analytical protocol adopted in NAA/CRNB laboratory [13].

Table 3. Example of typical ranges of experimental conditions of an INAA procedure.

Test portion mass : 5-500 mg			
Neutron fluence rates available $10^{16} - 10^{18}$ m^{-2} s^{-1}			
Irradiation	Decay	Measurement	Analyzed element
5 – 30 seconds	5 – 600 seconds	15 – 300 seconds	Short lived
1 – 8 hours	3 – 5 days	1 – 4 hours	Medium lived
	20 days	1 – 16 hours	Long lived

2.3. Derivation Of The Measurement Equation

The reaction rate R per nucleus capturing a neutron is given by:

$$R = \int_0^\infty \sigma(\upsilon)\phi'(\upsilon)d\upsilon = \int_0^\infty \sigma(E)\phi'(E)dE = \int_0^\infty n'(\upsilon)\upsilon\sigma(\upsilon)d\upsilon, \quad (1)$$

where:

σ (v) is the (n,γ) cross section (in cm^2; 1 barn (b) = 10^{-24} cm^2) at neutron velocity v (in cm s^{-1});

σ (E) is the (n,γ) cross section (in cm^2) at neutron energy E (in eV);

Φ'(v) is the neutron flux per unit of velocity interval (in cm^{-3}) at neutron velocity v;

n'(v) is the neutron density per unit of velocity interval (in cm^{-4} s) at neutron velocity v;

Φ'(E) is the neutron flux per unit of energy interval (in cm^{-2} s^{-1} eV^{-1}) at neutron energy E.

In Eq.(1), σ (v) = σ (E) with E (in erg = $6.2415.10^{11}$ eV) = ½ $m_n v^2$ [m_n rest mass of the neutron = 1.6749 10^{-24} g]. Furthermore, per definition, φ'(v) dv = φ'(E)dE (both in cm^{-2} s^{-1}).

In Eq.1, the functions σ(v) [= σ (E)] and φ'(v) [φ'(E)] are complex and are respectively depending on the (n,γ) reaction and on the irradiation site.

In 1987, F De Corte describes in his Aggregate thesis "Chapter 1: fundamentals [3] that the introduction of some generally valid characteristics yields the possibility of avoiding the actual integration and describing accurately the reaction rate in a relatively simple way by means of so-called formalisms or conventions. In short, these characteristics are:

In nuclear research reactors – which are intense sources of neutrons – three types of neutrons can be distinguished. The neutron flux distribution can be divided into three components (see Figure 9):

1. Fission or fast neutrons released in the fission of 235U. Their energy distribution ranges from 100 keV to 25 MeV with a maximum fraction at 2 MeV. These neutrons are slowed down by interaction with a moderator, e.g. H2O, to enhance the probability of them causing a fission chain reaction in the 235U.

2. The epithermal neutron component consists of neutrons (energies from 0.5 eV to about 100 keV). A cadmium foil 1 mm thick absorbs all thermal neutrons but will allow epithermal and fast neutrons above 0.5 eV in energy to pass through. Both thermal and epithermal neutrons induce (n,γ) reactions on target nuclei.

3. The thermal neutron component consists of low-energy neutrons (energies below 0.5 eV) in thermal equilibrium with atoms in the reactor's moderator. At room temperature, the energy spectrum of thermal neutrons is best described by a Maxwell-Boltzmann distribution with a mean energy of 0.025 eV and a most probable velocity of 2200 m/s. In general, a 1 MW reactor has a peak thermal neutron flux of approximately 10^{13} n/cm^2.

The (n,γ) cross section function, σ(v) versus v can be interpreted as a σ(v) ~ 1/v dependence, or σ (E) ~ $1/E^{1/2}$ dependence [log σ (E) versus log E is linear with slope -1/2], on which (above some eV) several resonances are superposed see Figure 10 taken from http://thorea.wikia.com/wiki/Thermal,_Epithermal_and_Fast_Neutron_Spectra web page.

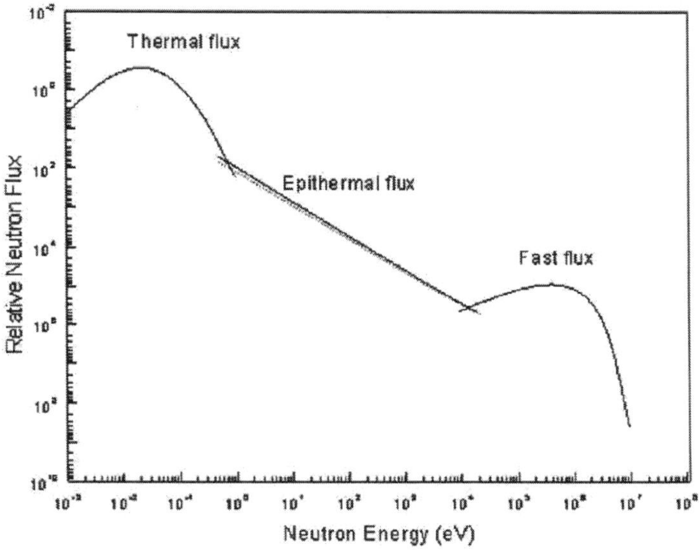

Figure 9. A typical reactor neutron energy spectrum showing the various components used to describe the neutron energy regions.

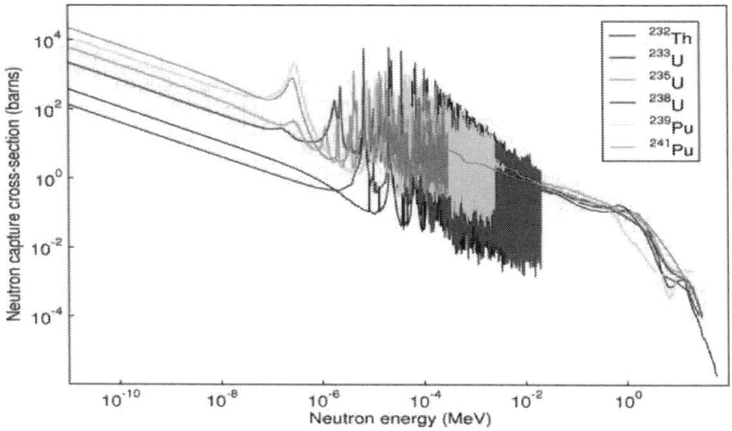

Figure 10. Relation between neutron cross section and neutron energy for major actinides (n, capture).

An NAA technique that employs only epithermal neutrons to induce (n,γ) reactions by irradiating the samples being analyzed inside either cadmium or boron a shield is called epithermal neutron activation analysis (ENAA).

The production of radioactive nuclei is described by:

$$\frac{dN}{dt} = RN_0 - \lambda N \tag{2}$$

In which N_0 number of target nuclei, N is the number of radioactive nuclei, λ is the decay constant in s^{-1}. The disintegration rate of the produced radionuclide at the end of the irradiation time ti follows from:

$$D(t_i) = N(t_i)\lambda = N_0 R(1 - e^{-\lambda t_i}) \tag{3}$$

where:

D is the disintegration rate in Bq of the produced radionuclide, assuming that N=0 at t=0 and N0=constant.

The dependence of the activation cross section and neutron fluence rate to the neutron energy can be taken into account in Eq. (1) by dividing the neutron spectrum into a thermal and an epithermal region; the division is made at En=0.55 eV (the so-called cadmium cut-off energy). This approach is commonly known as the Høgdahl convention [4].

The integral in Eq. (1) can then be rewritten as:

$$R = \int_0^{v_{Cd}} n(v)\,\mathrm{v}\,\sigma(v)dv + \int_{v_{Cd}}^{\infty} n(v)v\sigma(v)dv \tag{4}$$

The first term can be integrated straightforward:

$$\int_0^{v_{Cd}} n(v)vdv = v_0\sigma_0 \int_0^{v_{Cd}} n(v)dv = \mathrm{n}\mathrm{v}_0\sigma_0 \tag{5}$$

in which,

$$n = \int_{0}^{v_{Cd}} n(v)dv \tag{6}$$

is called the thermal neutron density, with $\Phi_{th}=nv_0$,

- Φ_{th} is the conventional thermal neutron fluence rate, $m^{-2}\,s^{-1}$, for energies up to the Cd cut-off energy of 0.55 eV;
- σ_0 is the thermal neutron activation cross section, m^2, at 0.025 eV;
- v_0 is the most probable neutron velocity at 20 °C: 2200 m s^{-1}.

The second term is re-formulated in terms of neutron energy rather than neutron velocity and the infinite dilution resonance integral I_0 – which effectively is also a cross section (m^2) – is introduced:

$$\int_{v_{Cd}}^{\infty} n(v)vdv = \varphi_{epi} \int_{E_{Cd}}^{E_{max}} \frac{\sigma(E_n)dE_n}{E_n} = \varphi_{epi} I_0 \tag{7}$$

with:

$$I_0 = \int_{E_{Cd}}^{E_{max}} \frac{\sigma(E_n)dE_n}{E_n} \tag{8}$$

Here, Φ_{epi} the conventional epithermal neutron fluence rate per unit energy interval, at 1 eV.

From this definition of I_0 it can be seen that it assumes that the energy dependency of the epithermal neutron fluence rate is proportional to 1/En. This requirement is fulfilled to a good approximation by most of the (n,γ) reactions.

In the practice of nuclear reactor facilities the epithermal neutron fluence rate Φepi is not precisely following the inverse proportionality to the neutron energy; the small deviation can be accounted for by introducing an epithermal fluence rate distribution parameter α:

$$I_0(\alpha) = (1\ \text{eV})^{\alpha} \int_{E_{Cd}}^{E_{max}} \frac{\sigma(E_n)dE_n}{E_n^{(1+\alpha)}} \tag{9}$$

The expression for the reaction rate can thus be re-written as:

$$R = \varphi_{th}\sigma_0 + \varphi_{epi}I_0(\alpha) \tag{10}$$

Expressing the ratio of the thermal neutron fluence rate and the epithermal neutron fluence rate as $f=\Phi_{th}/\Phi_{epi}$ and the ratio of the resonance integral and the thermal activation cross section as $Q_0(\alpha)= I_0(\alpha)/\sigma_0$, an effective cross section can be defined:

$$\sigma_{eff} = \sigma_0(1 + \frac{Q_0(\alpha)}{f}) \tag{11}$$

It simplifies the Eq. (10) for the reaction rate to:

$$R = \varphi_{th}\sigma_{eff} \tag{12}$$

This reaction rate applies to infinite thin objects. In objects of defined dimensions, the inside part will experience a lower neutron fluence rate than the outside part because neutrons are removed by absorption.

The nuclear transformations are established by measurement of the number of nuclear decays. The number of activated nuclei $N(t_i,t_d)$ present at the start of the measurement is given by:

$$N(t_i, t_d, t_m) = \frac{RN_0}{\lambda}(1 - e^{-\lambda t_i})e^{-\lambda t_d} \tag{13}$$

and the number of nuclei ΔN disintegrating during the measurement is given by:

$$\Delta N(t_i, t_d, t_m) = \frac{RN_0}{\lambda}(1 - e^{-\lambda t_i})e^{-\lambda t_d}(1\text{-}e^{-\lambda t_m}) \tag{14}$$

in which t_d is the decay or waiting time, i.e. the time between the end of the irradiation and the start of the measurement t_m is the duration of the measurement. Additional correction resulting from high counting rates may be necessary depending upon the gamma-ray spectrometer hardware used as illustrated in chapter 2 [1]. Replacing the number of target nuclei N_0 by $(N_{Av}m)/M$ and using the Eq. (12) for the reaction rate, the resulting net counts C in a peak in the spectrum corresponding with a given photon energy is approximated by the activation formula:

$$N_p = \Delta N\gamma\varepsilon = \varphi_{th}\sigma_{eff}\frac{N_{av}\theta m_x}{M_a}(1 - e^{-\lambda t_i})e^{-\lambda t_d}(1\text{-}e^{-\lambda t_m})I\varepsilon \tag{15}$$

with:

- N_p is the net counts in the γ-ray peak of E_γ;
- N_{Av} is the Avogadro's number in mol^{-1};
- θ is isotopic abundance of the target isotope;
- m_x is the mass of the irradiated element in g;
- M_a is the atomic mass in g mol^{-1};
- I is the gamma-ray abundance, i.e. the probability of the disintegrating nucleus emitting a photon of E_γ (photons $disintegration^{-1}$);
- ε is the full energy photopeak efficiency of the detector, i.e. the probability that an emitted photon of given energy will be detected and contribute to the photopeak at energy E_γ in the spectrum.

Although the photons emitted have energies ranging from tens of keVs to MeVs and have high penetrating powers, they still can be absorbed or scattered in the sample itself depending on the sample size, composition and photon energy. This effect is called gamma-ray self-attenuation.

Also, two or more photons may be detected simultaneously within the time resolution of the detector; this effect is called summation.

Eq. (15) can be simply rewritten towards the measurement equation of NAA, which shows how the mass of an element measured can be derived from the net peak area C:

$$m_x = N_p \cdot \frac{M_a}{N_{av} \cdot \theta} \cdot \frac{\lambda}{\varphi_{th} \cdot \sigma_{eff} \cdot \mathrm{I} \cdot \varepsilon \cdot (1 - e^{-\lambda t_i}) \cdot e^{-\lambda t_d} \cdot (1 - e^{-\lambda t_m})} \qquad (16)$$

2.4. Standardization

Standardization is based on the determination of the proportionality factors F that relate the net peak areas in the gamma-ray spectrum to the amounts of the elements present in the sample under given experimental conditions:

$$F = \frac{N_p}{M} \qquad (17)$$

Both absolute and relative methods of calibration exist.

2.4.1.absolute calibration

The values of the physical parameters determining the proportionality factor θ, N_{Av}, M, σ_{eff} I, λ, are taken from literature. The parameters σ_{eff} respectively I, λ are not precisely known for many (n,γ) reactions and radionuclides, and in some cases θ is also not accurately known. Since the various parameters were often achieved via independent methods, their individual uncertainties will add up in the combined uncertainty of measurement of the elemental amounts, leading to a relatively large combined standard uncertainty. Moreover, the metrological traceability of the values of the physical constants is not known for all radionuclides. The other parameters N_p, m_x, Φ, ε, t_i, t_d, t_m are determined, calculated or measured for the given circumstances and uncertainties can be established.

2.4.2. Relative calibration

a. Direct comparator method

The unknown sample is irradiated together with a calibrator containing a known amount of the element(s) of interest. The calibrator is measured under the same conditions as the sample (sample-to-detector distance, equivalent sample size and if possible equivalent in composition). From comparison of the net peak areas in the two measured spectra the mass of the element of interest can be calculated:

$$m_{x(unk)} = m_{x(cal)} \cdot \frac{\left(\dfrac{N_p}{t_m . e^{-\lambda t_d} . (1 - e^{-\lambda t_m})}\right)_{unk}}{\left(\dfrac{N_p}{t_m . e^{-\lambda t_d} . (1 - e^{-\lambda t_m})}\right)_{cal}} \tag{18}$$

in which m_x(unk), m_x(cal) mass of the element of interest, in the unknown sample and the calibrator, respectively in g.

In this procedure many of the experimental parameters - such as neutron fluence rate, cross section and photopeak efficiency cancel out at the calculation of the mass and the remaining parameters are all known. This calibration procedure is used if the highest degree of accuracy is required.

The relative calibration on basis of element calibrators is not immediately suitable for laboratories aiming at the full multi-element powers of INAA. It takes considerable effort to prepare multi-element calibrators for all 70 elements measurable via NAA with adequate degree of accuracy in a volume closely matching the size and the shape of the samples. Single comparator method Multi-element INAA on basis of the relative calibration method is feasible when performed according to the principles of the single comparator method. Assuming stability in time of all relevant experimental conditions, calibrators for all elements are co-irradiated each in turn with the chosen single comparator element. Once the sensitivity for all elements relative to the comparator element has been determined (expressed as the so-called k-factor, see below), only the comparator element has to be used in routine measurements instead of individual calibrators for each element. The single comparator method for multi-element INAA was based on the ratio of proportionality factors of the element of interest and of the comparator element after correction for saturation, decay, counting and sample weights defined the k-factor for each element i as:

$$k_i = \frac{(M_a)_{i,cal}\gamma_{comp}\varepsilon_{comp}\theta_{comp}\sigma_{eff,comp}}{M_{i,cal}\gamma_{i,cal}\varepsilon_{i,cal}\theta_{i,cal}(\sigma_{eff})_{i,cal}} \tag{19}$$

Masses for each element i then can be calculated from these k_i factors; for an element determined via a directly produced radionuclide the mass m_x(unk) follows from:

$$m_{x(unk)} = m_{x(comp)} \cdot \frac{\left(\dfrac{N_p}{(1-e^{-\lambda t_m}).t_m.e^{-\lambda t_d}.(1-e^{-\lambda t_m})}\right)_{unk}}{\left(\dfrac{N_p}{(1-e^{-\lambda t_m}).t_m.e^{-\lambda t_d}.(1-e^{-\lambda t_m})}\right)_{comp}} .k_i \tag{20}$$

where: m_x(comp) is the mass of element x in comparator in g.

These experimentally determined k-factors are often more accurate than when calculated on basis of literature data as in the absolute calibration method. However, the k-factors are only valid for a specific detector, a specific counting geometry and irradiation facility, and remain valid only as long as the neutron fluence rate parameters of the irradiation facility remain stable. The single comparator method requires laborious calibrations in advance, and finally yield relatively (compared to the direct comparator method) higher uncertainties of the measured values. Moreover, it requires experimental determination of the photopeak efficiencies of the detector. Metrological traceability of the measured values to the S.I. may be demonstrated.

a. The k_0-comparator method

The k_0-based neutron activation analysis (k_0-NAA) technique, developed in 1970s, is being increasingly used for multielement analysis in a variety of matrices using reactor neutrons [4-10]. In our research reactor, the k_0-method was successfully developed using the Høgdahl formalism [11]. In the $k0$-based neutron activation analysis the evaluation of the analytical result is based on the so-calledk_0- factors that are associated with each gamma-line in the gamma-spectrum of the activated sample. These factors replace nuclear constants, such as cross sections and gamma-emission probabilities, and are determined in specialized NAA laboratories. This technique has been reported to be flexible with respect to changes in irradiation and measuring conditions, to be simpler than the relative comparator technique in terms of experiments but involves more complex formulae and calculations, and to eliminate the need for using multielement standards. The k_0-NAA technique, in general, uses input parameters such as (1) the epithermal neutron flux shape factor (α), (2) subcadmium-to-epithermal neutron flux ratio (f), (3) modified spectral index

$r(\alpha)\sqrt{T_n / T_0}$, (4) Westcott's $g(Tn)$-factor, (5) the full energy peak detection efficiency (ε_p), and (6) nuclear data on Q_0 (ratio of resonance integral (I_0) to thermal neutron cross section (σ_0) and k_0. The parameters from (1) to (4) are dependent on each irradiation facility and the parameter (5) is dependent on each counting facility. The neutron field in a nuclear reactor contains an epithermal component that contributes to the sample neutron activation [12]. Furthermore, for nuclides with the Westcott's $g(Tn)$-factor different from unity, the Høgdahl convention should not be applied and the neutron temperature should be introduced for application of a more sophisticated formalism [14], the Westcott formalism. These two formalisms should be taken into account in order to preserve the accuracy of k_0-method.

The k_0-NAA method is at present capable of tackling a large variety of analytical problems when it comes to the multi-element determination in many practical samples. In this part, we have published a paper [15] for which the determination of the Westcott and Høgdahl parameters have been carried out to assess the applicability of the k_0-NAA method using the experimental system and irradiation channels at Es-Salam research reactor.

During the three last decades Frans de Corte and his co-workers focused their investigations to develop a method based on co-irradiation of a sample and a neutron flux monitor, such as gold and the use of a composite nuclear constant called k_0-factor [3, 16]. In addition, this method allows to analyze the sample without use the reference standard like INAA method. The k-factors have been defined as independent of neutron fluence rate parameters as well as of spectrometer characteristics. In this approach, the irradiation parameter $(1+Q_0(\alpha)/f)$ (Eq. (11)) and the detection efficiency ε are separated in the expression (19) of the k-factor, which resulted at the definition of the k_0-factor.

$$k_0 = \frac{1}{k} \cdot \frac{1+Q_{0,\mathrm{comp}}(\alpha)/f}{1+Q_{0,\mathrm{cal}}(\alpha)/f} \cdot \frac{\varepsilon_{\mathrm{comp}}}{\varepsilon_{\mathrm{comp}}} = \frac{M_{\mathrm{comp}}}{\theta_{comp}\sigma_{0,comp}\gamma_{comp}} \cdot \frac{\theta_{cal}\sigma_{0,cal}\gamma_{cal}}{M_{\mathrm{cal}}} \tag{21}$$

$$m_{x(unk)} = m_{x(comp)} \frac{1+Q_{0,\mathrm{comp}}(\alpha)/f}{1+Q_{0,\mathrm{cal}}(\alpha)/f} \cdot \frac{\varepsilon_{\mathrm{comp}}}{\varepsilon_{\mathrm{comp}}} \cdot \frac{\left(\dfrac{N_p/t_m}{(1-e^{-\lambda t_i}).e^{-\lambda t_d}(1-e^{-\lambda t_m}).m}\right)_{\mathrm{unk}}}{\left(\dfrac{N_p/t_m}{(1-e^{-\lambda t_i}).e^{-\lambda t_d}(1-e^{-\lambda t_m}).m}\right)_{comp}} \cdot \frac{1}{k_0} \tag{22}$$

The applicability of HØGDAHL convention is restricted to (n,γ) reactions for which WESTCOTT's g-factor is equal to unity (independent of neutron temperature), the cases for which WESTCOTT's g = 1 [3, 4, 17], such as the reactions $^{151}Eu(n, \gamma)$ and $^{176}Lu(n, \gamma)$ are excluded from being dealt with. Compared with relative method k_0-NAA is experimentally simpler (it eliminates the need for multi-element standards [3, 18], but requires more complicated calculations [19]. In our research reactor, the k_0-method was successfully developed using the HØGDAHL convention and WESTCOTT formalism [11, 15]. The k_0-method requires tedious characterizations of the irradiation and measurement conditions and results, like the single comparator method, in relatively high uncertainties of the measured values of the masses. Moreover, metrological traceability of the currently existing k_0 values and associated parameters to the S.I. is not yet transparent and most probably not possible. Summarizing, relative calibration by the direct comparator method renders the lowest uncertainties of the measured values whereas metrological traceability of these values to the S.I. can easily be demonstrated. As such, this approach is often preferred from a metrological viewpoint. The concentration of an element can be determined as:

$$\rho_x(ppm) = \frac{\left[\frac{N_p/t_m}{SDCW}\right]_x}{\left[\frac{N_p/t_m}{SDCW}\right]_{Au}} \cdot \frac{1}{k_{0,Au(x)}} \cdot \frac{G_{th,Au} f + G_{epi,Au} Q_{0,Au}(\alpha)}{G_{th,x} f + G_{epi,x} Q_{0,x}(\alpha)} \cdot \frac{\varepsilon_{p,Au}}{\varepsilon_{p,x}} \times 10^6 \tag{23}$$

Where: the indices x and Au refer to the sample and the monitor, respectively; W_{Au} and W_x represent the mass of the gold monitor and the sample (in g); N_p is the measured peak area, corrected for dead time and true coincidence; S, D, C are the saturation, decay and counting factors, respectively; tm is the measuring time; G_{th} and G_e are the correction factors for thermal and epithermal neutron self shielding, respectively.

2.5. Sources errors

Many publications reported in literature [20-25] treat the concept of evaluation of uncertainties in large range of analytical techniques.

We can give in this part of chapter, the evaluation of uncertainties for neutron activation analysis measurements. Among the techniques of standardization the comparator method for which the individual uncertainty components associated with measurements made with neutron activation analysis (NAA) using the comparator method of standardization (calibration), as well as methods to evaluate each one of these uncertainty components [1].

This description assumes basic knowledge of the NAA method, and that experimental parameters including sample and standard masses, as well as activation, decay, and counting times have been optimized for each measurement. It also assumes that the neutron irradiation facilities and gamma-ray spectrometry systems have been characterized and optimized appropriately, and that the choice of irradiation facility and detection system is appropriate for the measurement performed. Careful and thoughtful experimental design is often the best means of reducing uncertainties. The comparator method involves irradiating and counting a known amount of each element under investigation using the same or very similar conditions as used for the unknown samples. Summarizing, relative calibration by the direct comparator method renders the lowest uncertainties of the measured values whereas metrological traceability of these values to the S.I. can easily be demonstrated. As such, this approach is often preferred from a metrological viewpoint. The measurement equation can be further simplified, by substituting:

$$A_0 = \frac{\lambda N_p e^{\lambda t_d}}{(1-e^{-\lambda t_m})\cdot(1-e^{-\lambda t_i})}.f_p.f_{ltc} \tag{24}$$

in :

$$m_{unk} = m_{cal}\frac{A_{0(unk)}}{A_{0(cal)}}R_\theta R_\Phi R_\sigma R_\varepsilon\text{-blank} \tag{25}$$

Where: R_θ is the ratio of isotopic abundances for unknown sample and calibrator, R_ϕ is the ratio of neutron fluence rates (including fluence gradient, neutron self shielding, and scattering) for unknown sample and calibrator, R_σ is the ratio of effective cross sections if neutron spectrum shape differs from unknown sample to calibrator, R_ε is the ratio of counting efficiencies (differences due to geometry and γ-ray self shielding) for unknown sample and calibrator, blank is the mass of element x in the blank, $f_p$is the correction for pulse pileup (correction method depends upon the actual hardware used) and f_{ltc} is the correction for inadequacy of live time extension (correction method depends upon the actual hardware used)

Note that the R values are normally very close to unity, and all units are either SI-based or dimensionless ratios. Thus an uncertainty budget can be developed using only SI units and dimensionless ratios for an NAA measurement by evaluating the uncertainties for each of the terms inEqs. (23) and (24), and for any additional corrections required (e.g., interferences, dry mass conversion factors, etc.).

Uncertainties for some of the terms in Eq. (24) have multiple components. If we sub-divide the uncertainty for each term in the above equations into individual components, add terms for potential corrections, and separate into the four stages of the measurement process, including: pre-irradiation (sample preparation); irradiation; post-irradiation (gamma-ray spectrometry), and radiochemistry, we arrive at the complete list of individual uncertainty components for NAA listed below in Table 4. Only uncertainties from the first three stages should be considered for instrumental neutron activation analysis (INAA) measurements, while all four stages should be considered for radiochemical neutron activation analysis (RNAA) measurements. More details are given in chapter 2 of reference [1] for each subsection of uncertainty component.

Table 4. Complete list of individual uncertainty components for NAA measurements using the comparator method of standardization; line numbers in this table represent subsections.

1. Pre-irradiation (sample and standard preparation) stage
1.1. Elemental content of standards (comparators)
1.2. Target isotope abundance ratio — unknown samples/standards
1.3. Basis mass (or other sample basis) — including drying
1.4. Sample and standard blanks
2. Irradiation stage
2.1. Neutron fluence exposure differences (ratios) for unknown samples compared to standards (comparators)
2.1. Physical effects (fluence gradients within a single irradiation)
2.2. Temporal effects (fluence variations with time)
2.3. Neutron self shielding (absorption and scattering) effects within a single sample or standard
2.4. Neutron shielding effects from neighbouring samples or standards

2.2. Irradiation interferences
2.2.1. Fast (high energy) neutron interferences
2.2.2. Fission interferences
2.2.3. Multiple neutron capture interferences
2.3. Effective cross section differences between samples and standards
2.4. Irradiation losses and gains
2.4.1. Hot atom transfer (losses and gains by recoil, nanometer movement)
2.4.2. Transfer of material through irradiation container
2.4.3. Sample loss during transfer from irradiation container
2.4.4. Target isotope burn up differences
2.5. Irradiation timing and decay corrections during irradiation (effects of half life and timing uncertainties)
3. Gamma-ray spectrometry stage
3.1. Measurement replication or counting statistics (depending on number of replicates) for unknown samples
3.2. Measurement replication or counting statistics (depending on number of replicates) for comparator standards
3.3. Corrections for radioactive decay (effects of half life and timing uncertainties for each measurement)
3.3.1. From end of irradiation to start of measurement
3.3.2. Effects of clock time uncertainty
3.3.3. Effects of live time uncertainty
3.3.4. Count-rate effects for each measurement
3.3.4.1. Corrections for losses due to pulse pileup for conventional analyzer systems
3.3.4.2. Effects due to inadequacy of live-time extension for conventional analyzer systems
3.3.4.3. Uncertainties due to hardware corrections for Loss-Free or Zero Dead Time systems
3.4. Corrections for gamma-ray interferences
3.5. Corrections for counting efficiency differences (if necessary), or uncertainty for potential differences
3.5.1. Effects resulting from physical differences in size and shape of samples versus standards
3.5.2. Corrections for gamma-ray self absorption
3.6. Potential bias due to peak integration method
3.7. Potential bias due to perturbed angular correlations (-ray directional effects)
4. Radiochemical stage (only if radiochemical separations are employed)
4.1. Losses during chemical separation
4.2. Losses before equilibration with carrier or tracer

2.6. Detection Limits Of Naa

The detection limit represents the ability of a given NAA procedure to determine the minimum amounts of an element reliably. The detection limit depends on the irradiation, the decay and the counting conditions.

It also depends on the interference situation including such things as the ambient background, the Compton continuum from higher energy-rays, as well as any-ray spectrum interferences from such factors as the blank from pre-irradiation treatment and from packing materials. The detection limit is often calculated using Currie's formula:

$$DL = 2.71 + 4.65B, \tag{26}$$

where: DL is the detection limit and B is the background under a gamma-ray peak. This relation is valid only when the gamma-ray background (counting statistical error) is the major interference.

However, practically, the INAA detection limits depend on:

1. The amount of material to be irradiated and to be counted. This is often set by availability, sample encapsulation aspects and safety limits both related to irradiation (irradiation containers) and counting (e.g. with Ge well-type detectors), and possibly because of neutron self-shielding and gamma-ray self-absorption effects. For these reasons practically the sample mass is often limited to approximately 250 mg.

2. The neutron fluxes. These are clearly set by the available irradiation facilities.

3. The duration of the irradiation time. This is set by practical aspects, such as the limitations in total irradiation dose of the plastic containers because of radiation damage. The maximum irradiation time for polyethylene capsules is usually limited to several hours, for instance 5 hours at 5×10^{17} $m^{-2}s^{-1}$.

4. The total induced radioactivity that can be measured is set by the state-of-the-art of counting and signal processing equipment, with additional radiation dose and shielding considerations. As an example, the maximum activity at the moment of counting may have to be limited to approximately 250 kBq.

5. The duration of the counting time. A very long counting time may set limits to the number of samples processed

simultaneously in case the radioactivity decays considerably during this counting time. Moreover, it reduces sample throughput.

6. The total turn-around time. Although sometimes better detection limits may be obtained at long decay times, the demands regarding the turn-around time often imply that a compromise has to be found between the longest permissible decay time and customer satisfaction.

7. The detector size, counting geometry and background shielding. The detector's characteristics may be set in advance by availability but several options exist.

It all illustrates that the detection limit for a given element by INAA may be different for each individual type of material, and analysis conditions. In Table 5 are given, as an indication, typical detection limits as derived from the analysis of a plant and a soil material. Peter Bode in his PhD thesis, Instrumental and organizational aspects of a neutron activation analysis laboratory, the typical detection limits as derived from the analysis of a plant and a soil material given in table 5 [26].

Table 5. Detection limits of elements in $mg.kg^{-1}$ as observed in NAA procedure of plant material and a soil material.

Element	Plant	Soil	Element	Plant	Soil	Element	Plant	Soil
NA	2	10	Nd	0.7	8	Ag	0.2	2
Ca	700	4000	Eu	0.006	0.05	Sn	10	20
Cr	1	1	Yb	0.03	0.2	Te	0.3	3
Co	0.02	0.3	Hf	0.01	0.1	Ba	10	40
Zn	0.4	6	W	0.3	1	Ce	0.2	1
As	0.2	0.8	Os	0.1	0.6	Sm	0.01	0.03
Br	0.3	0.8	Au	0.003	0.01	Tb	0.008	0.1
Sr	5	60	Th	0.01	0.1	Lu	0.004	0.02
Mo	4	10	K	200	1500	Ta	0.01	0.2
Cd	3	8	Sc	0.001	0.02	Re	0.08	0.2
Sb	0.02	0.2	Fe	8	100	Ir	0.0006	0.004
Cs	0.02	0.3	Ni	2	30	Hg	0.05	0.4
La	0.1	0.3	Ga	2	10	U	0.2	2
Se	0.1	1	Rb	0.4	6	Zr	5	80

3. APPLICATIONS

It is hardly possible to provide a complete survey of current NAA applications; however, some trends can be identified [27]. At specialized institutions, NAA is widely used for analysis of samples within environmental specimen banking programmes [28]. The extensive use of NAA in environmental control and monitoring can be demonstrated by the large number of papers presented at two symposia organized by the IAEA in these fields: "Applications of Isotopes and Radiation in Conservation of the Environment" in 1992 [29] and "Harmonization of Health-Related Environmental Measurements Using Nuclear and Isotopic Techniques" in 1996 [30]. Similar trends can also be identified from the topics discussed at the regular conference on "Modern Trends in Activation Analysis (MTAA)" and at the symposia on "Nuclear Analytical Methods in the Life Sciences" [31-33]. Additional sources of recent information on utilizing NAA in selected fields, such as air pollution and environmental analysis, food, forensic science, geological and inorganic materials as well as water analysis can be found in the bi-annual reviews in Analytical Chemistry, for instance cf. Refs [34-42]. It follows from these reviews that NAA has been applied for determining many elements, usually trace elements, in the following fields and sample types:

1. Archaeology: samples and objects such as amber, bone, ceramics, coins, glasses, jewellery, metal artefacts and sculptures, mortars, paintings, pigments, pottery, raw materials, soils and clays, stone artefacts and sculptures can be easily analyzed by NAA.

2. Biomedicine: the samples and objects that can be analysed include: animal and human tissues activable tracers, bile, blood and blood components, bone, brain cell components and other tissues, breast tissue, cancerous tissues, colon, dialysis fluids, drugs and medicines, eye, faeces, foetus, gallstones, hair, implant corrosion, kidney and kidney stones, liver, lung, medical plants and herbs, milk, mineral availability, muscle, nails, placenta, snake venom, rat tissues (normal and diseased), teeth, dental enamel and dental fillings, thyroid, urine and urinary stones.

In this work, we have used the INAA technique to analyse the traditional medicinal seeds prescribed for specific treatment purposes, were purchased from local markets [43]. The samples were irradiated at Es-Salam research reactor, at a power of 5 MW for 6 h. The accuracy of the method was established by analyzing reference materials. Twenty elements were measured, with good accuracy and reproducibility (Table 6[1] -). The concentration of elements determined, was found to vary depending on the seeds (Fig.11). The daily intake of essential and toxic elements was determined, and compared with the recommended values. The probable cumulative intake of toxic elements is well below the tolerance limits.

Table 6. Elemental concentrations in the medicinal seed samples (Black seeds, Fenugreek, Caraway).

Element	Unity	Black seeds	Fenugeek	Caraway
Ba	mg/Kg	7.7 ± 5.5	100.3 ± 5.8	112.4 ± 6.5
Br	mg/Kg	136.9 ± 4.6	119.6 ± 3.9	72.9 ± 2.4
Ca	g/Kg	3.77 ± 4.55	3.14 ± 0.39	1.50 ± 0.21
Ce	mg/Kg	1.44± 0.07	2.6 ± 0.1	1.98 ± 0.11
Co	mg/Kg	0.66 ± 0.02	0.73 ± 0.02	0.81 ± 0.03
Cr	mg/Kg	4.44 ± 0.19	29.3 ± 1.0	2.96 ± 0.20
Cs	mg/Kg	0.25 ± 0.01	0.51 ± 0.02	0.22 ± 0.01
Eu	mg/Kg	0.022 ± 0.002	0.039 ± 0.002	0.023 ± 0.002
Fe	mg/Kg	656.2 ± 71.6	823.2 ± 89.8	674.67 ± 74.16
K	g/Kg	3.67± 1.79	3.75 ± 0.20	3.7 ± 0.2
La	mg/Kg	0.74 ± 0.04	1.53 ± 0.06	1.50 ± 0.06
Na	mg/Kg	1028 ± 34	804.20 ± 26.69	615.50 ± 20.41
Rb	mg/Kg	24 ± 2	36.8 ± 1.4	26.3 ± 1.9
Sc	mg/Kg	0,258 ± 0,037	0,362 ± 0,051	0,272 ± 0,008
Se	mg/Kg	0,29 ± 0,04	ND	ND
Sm	mg/Kg	0,092 ± 0,004	0,18 ± 0,01	0,142 ± 0,005
Sr	mg/Kg	203,2 ± 7,8	136,88 ± 7,4	101,7 ± 4,7
Th	mg/Kg	0,159 ± 0,009	0,32 ± 0,02	0,195 ± 0,014
Zn	mg/Kg	68,06 ± 2,11	42,8 ± 1,4	40,24 ± 1,30

1 nd : not detected.

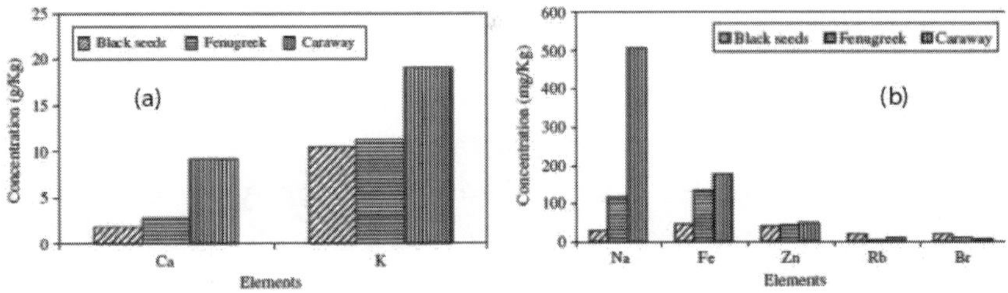

Figure 11. Concentration of the major and minor (a) and trace (b) elements in the medicinal seed samples.

1. Environmental: in this domain, related fields concerned by NAA are: aerosols, atmospheric particulates (size fractionated), dust, fossil fuels and their ashes, flue gas, animals, birds, insects, fish, aquatic and marine biota, seaweed, algae, lichens, mosses, plants, trees (leaves, needles, tree bark), household and municipal waste, rain and horizontal precipitations (fog, icing, hoarfrost), soils, sediments and their leachates, sewage sludges, tobacco and tobacco smoke, surface and ground waters, volcanic gases.

Recently, our laboratory is strongly involved in various areas of application of k_0-NAA. The present work focuses on the application of the k_0-NAA method in Nutritional and Health-Related Environmental field [44]. Tobacco holds a leading position among different commodities of human consumption. The adverse health effects of toxic and trace elements in tobacco smoke on smokers and non-smokers are a special concern. In the present study, the concentration of 24 trace elements in cigarette tobacco of five different brands of Algerian and American cigarettes have been determined by k_0-based INAA method. The results were compared with those obtained for samples from Iranian, Turkish, Brazilian and Mexican cigarettes tobacco. To evaluate the accurate of the results the SRM IAEA-140/TM was executed.

A multi-element analysis procedure based on the k_0-NAA method was developed at Es-Salam research reactor allowing to simultaneously determine concentrations for 24 elements (As, Ba, Br, Ca, Ce, Co, Cr, Cs,

Eu, Fe, Hf, K, La, Na, Rb, Sb, Sc, Se, Sm, Sr, Ta, Tb, Th, Zn). The determination of toxic and trace elements in cigarette tobacco is important both from the point of view of health studies connected with smoking and more general aspects of the uptake of trace elements by plants (table 7). Because of its great sensitivity, k_0-NAA method is very suitable for determination of heavy metals such as As, Sb, Se and Zn. The accuracy of the results was checked by the analysis of standard reference material and good agreement was obtained with certified or literature values. The results of Algerian tobacco (table 8) were compared with analyses of Turkey [45], Iran [46], Mexican [47] and Brazilian tobacco [48].

Table 7. Concentration values (mg kg-1) of five brands of tobacco by k_0-NAA method.

Element	Designated sample code				
	T1	T2	T3	T4	T5
Sr	203.2 ± 7.8	136.9 ± 7.4	101.7 ± 4.7	82.37 ± 3.62	106.80 ± 5.39
Ta	0.021 ± 0.003	0.043 ± 0.006	0.023 ± 0.004	0.029 ± 0.004	0.023 ± 0.004
Tb	0.018 ± 0.004	0.021 ± 0.003	0.014 ± 0.004	0.008 ± 0.002	0.018 ± 0.004
Th	0.159 ± 0.009	0.32 ± 0.02	0.195 ± 0.014	0.153 ± 0.012	0.19 ± 0.02
Zn	68.06 ± 2.11	42.8 ± 1.4	40.24 ± 1.30	27.53 ± 0.89	42.99 ± 1.38

Table 8. Comparison between our results (Algerian cigarettes tobacco) and those reported in the literature.

Element	Algeria	Turkey	Iran	Mexican	Brazilian
As	4.05 – 6.4	1.0	-	< 0.55 – 3.24	-
Ba	100.3 – 101.3	64.6	1.15 ± 0.01	64 – 251	45.8 – 99.7
Br	119.6 – 136.9	59.2	137 ± 2	49 -136	-
Ca %	3.14 – 3.77	-	-	-	-
Ce	1.44 – 2.6	-	1.19 ± 0.02	< 0.3 -1.7	1.2 – 8.3
Co	0.66 – 0.73	0.77	0.21 ± 0.01	0.29 – 0.55	0.70 – 1.2
Cr	4.44 – 29.3	7.1	3.14 ± 0.14	< 0.8 – 2.4	-
Cs	0.25 – 0.51	0.20	0.17 ± 0.02	0.091 – 0.4	-
Eu	0.022 – 0.039	0.034	0.045 ± 0.010	< 0.03	0.036 -0.037
Fe	656.2 – 823.2	1000	649 ± 6	420 – 680	710 – 4100
Hf	0.127 – 0.224	0.15	0.09 ± 0.01	< 0.03 – 0.13	0.15 – 0.87
K %	3.67 – 3.75	2.6	0.220 ± 0.005	1.83 – 4.03	-
La	0.74 – 1.53	1.0	0.62 ± 0.02	< 0.2 – 0.66	1.9 – 4.8
Na	804.2 – 1028	394	347 ± 18	309 – 566	-
Rb	24.0 - 36.8	17.7	22.6 ± 3.6	19 – 50	30.3 – 45.0
Sb	0.089 – 0.180	0.10	0.68 ± 0.03	< 0.7	0.079 – 0.47
Sc	0.258 – 0.362	0.35	0.43 ± 0.03	0.13 – 0.22	0.23 – 1.50

1. Forensics: bomb debris, bullet lead, explosives detection, glass fragments, paint, hair, gunshot residue swabs, shotgun pellets.

2. Geology and geochemistry: asbestos, bore hole samples, bulk coals and coal products, coal and oil shale components, crude oils, kerosene, petroleum, cosmo-chemical samples, cosmic dust, lunar samples, coral, diamonds, exploration and geochemistry, meteorites, ocean nodules, rocks, sediments, soils, glacial till, ores and separated minerals.

3. Industrial products: alloys, catalysts, ceramics and refractory materials, coatings, electronic materials, fertilizers, fissile material detection and other safeguard materials, graphite, high purity and high-tech materials, integrated circuit packing materials, online, flow analysis, oil products and solvents, pharmaceutical products, plastics, process control applications, semiconductors, pure silicon and silicon processing, silicon dioxide, NAA irradiation vials, textile dyes, thin metal layers on various substrates.

4. Nutrition: composite diets, foods, food colours, grains, honey, seeds, spices, vegetables, milk and milk formulae, yeast. In this chapter, we focus on the application of the k_0 method of instrumental neutron activation analysis in Nutritional and Health-Related Environmental field [49]. Three kinds of milk were purchased in the powder form from local supermarket. The samples of milk powder were analyzed using k_0-NAA method. Concentrations of six elements Br, Ca, K, Na, Rb and Zn have been determined by long irradiation time with a thermal and epithermal flux of $4.7.10^{12}$ $n.cm^{-2}.s^{-1}$ and $2.29.10^{11}$ $n.cm^{-2}.s^{-1}$, respectively (see table 9). The reactor neutron spectrum and detection efficiency calibration parameters such as α, f and εp have been used for the calculation of elemental concentrations. The analytical results for three kinds of milk using k_0-NAA are compared with the certified values of SRMs. In this work, we have determined six elements in three kinds of milk and two reference materials, IAEA-153 and IAEA-155. The elements Br, Ca, K, Na, Rb and Zn were determined in each kind of the three samples of milk.

Table 9. Concentration values of Milk: M1, M2 and M3; units are in mg/kg, NB: (value) is the concentration value of indicated by producer.

Element	Designated sample code				
	T1	T2	T3	T4	T5
As	4.05 ± 0.16	6.4 ± 0.24	2.16 ± 0.09	2.42 ± 0,09	4.27 ± 0.28
Ba	101.3 ± 5.5	100.3 ± 5.8	112.4 ± 6.5	83.1 ± 4.4	120.0 ± 7.7
Br	136.9 ± 4.6	119.6 ± 3.9	72.9 ± 2.4	54.1 ± 1.7	68.7 ± 2.3
Ca %	3.77 ± 4.55	3.14 ± 0.39	1.50 ± 0.21	2.11 ± 0.26	2.39 ± 0.31
Ce	1.44 ± 0.07	2.6 ± 0.1	1.98 ± 0.11	1.01 ± 0.05	1.81 ± 0.11
Co	0.66 ± 0.02	0.73 ± 0.02	0.81 ± 0.03	0.51 ± 0.02	0.78 ± 0.03
Cr	4.44 ± 0.19	29.3 ± 1.0	2.96 ± 0.20	2.37 ± 0.11	2.80 ± 0.22
Cs	0.25 ± 0.01	0.51 ± 0.02	0.22 ± 0.01	0.191 ± 0.008	0.42 ± 0.02
Eu	0.022 ± 0.002	0.039 ± 0.002	0.023 ± 0.002	0.021 ± 0.001	0.032 ± 0.002
Fe	656 ± 72	823 ± 90	675 ± 74	384 ± 13	664 ± 24
Hf	0.127 ± 0.007	0.224 ± 0.015	0.24 ± 0.02	0.144 ± 0.014	0.143 ± 0.010
K %	3.67± 1.79	3.75 ± 0.20	3.7 ± 0.2	2.92 ± 0.14	3.38 ± 0 .18
La	0.74 ± 0.04	1.53 ± 0.06	1.50 ± 0.06	1.082 ± 0.043	1.68 ± 0.06
Na	1028 ± 34	804 ± 27	616 ± 20	653 ± 22	575 ± 19
Rb	24 ± 2	36.8 ± 1.4	26.3 ± 1.9	14.33 ± 0.59	25.4 ± 1.6
Sb	0.089 ± 0.014	0.180 ± 0.020	0.036 ± 0.007	0.127 ± 0.015	0.346 ± 0.037
Sc	0.258 ± 0.037	0.362 ± 0.051	0.272 ± 0.008	0.165 ± 0.023	0.264 ± 0.037
Se	0.29 ± 0.04	ND	ND	ND	ND
Sm	0.092 ± 0.004	0.18 ± 0.01	0.142 ± 0.005	0.095 ± 0.004	0.152 ± 0.005

The accuracy of the measurements was evaluated by analyzing two SRMs Whey powder AIEA-155 and Milk powder AIEA-153. The analysis results illustrated in figure 12 showed that the deviations between experimental and certified values were mostly less than 10%.

As an example, an investigation in the nutrition field was carried out by the radiochemical neutron activation analysis to the proportioning of iodine in food salt [50].

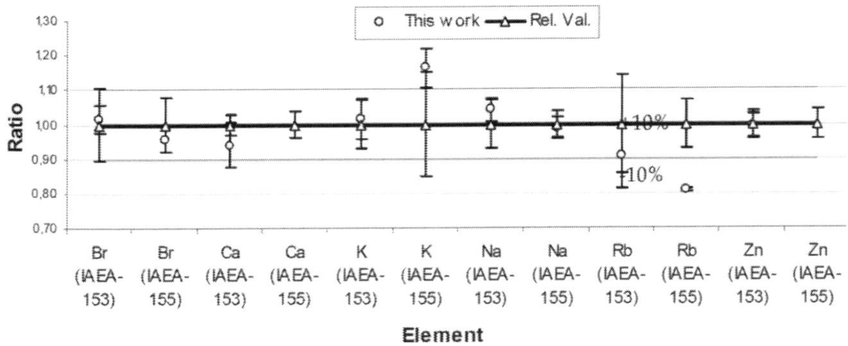

Figure 12. Comparison of k_0-NAA data to certified values for IAEA-153 and IAEA-155.

1. Quality assurance: this include analysis of reference materials, certification of element contents and homogeneity testing of mainly biological and environmental reference materials of chemical composition and methods inter-comparisons. Additional information about these applications can be found in the Proceedings of the Int. Symposia on Biological and Environmental Reference Materials (BERM). In 2012, Hamidatou L et all reported "k0-NAA quality assessment in an Algerian laboratory by analysis of SMELS and four IAEA reference materials using Es-Salam Research reactor" the internal quality control of the k0-NAA technique [51]. The concept of QC/QA, internal and external validation is considered as an advanced stage in the life cycle of an analytical method.

Our contribution in this domain is considered as periodic activities. Since the Nineties our laboratory participated through AFRA/AIEA projects in different inter-laboratory proficiency tests. Recently, our laboratory was participated in four inter-comparison tests organized by IAEA within the framework of the AFRA project to assess the analytical performance of 18 analytical laboratories participating in the RAF /4/022 project, Enhancement of Research Reactor Utilization and Safety by taking part in analytical proficiency testing IAEA in conjunction with WEPAL, the Wageningen Evaluating Programs for Analytical Laboratories. The Proficiency Testing tests related to the determination of major, minor and trace elements in materials of the International Soil and Plant Analytical Exchange material (Wepal codes ISE, IPE).

1. Neutron flux characterization: theoretical and experimental study, calibration of irradiation channels, simulations using Monte Carlo Code. In general, the implementation of new techniques based on the neutron beams or flux around the research reactors needs the knowledge of the essential parameters of neutron flux in different sites to obtain a better precision during the development.

In this context, we give a great interest in the neutron study for our irradiation channels by making periodic calibrations using experimental and simulation approaches [11, 15, 52].

4. CONCLUSIONS

NAA plays a complementary role in materials analysis in an industrial analytical laboratory. There are applications where it is highly desirable, and may play the dominant role as the method of choice e.g. bulk analysis of Si. The advantages of NAA are still the minimum sample preparation and ultra high sensitivity while turnaround time and lack of spatial resolution is a significant limitation.

The many diverse applications in varied fields show that NAA is extremely useful, even though it is a relatively simple analytical method. Irradiation of samples may be done at nuclear reactors that offer such services. Such centres are easily accessible nowadays. Even if these centres are inaccessible, other neutron sources that emit thermal neutrons may be used. Hence any laboratory that has a gamma counter can perform NAA experiments. Development of research programs based on NAA is accessible to any laboratory that is willing to invest a minimal amount of funds. In this chapter, we have presented the neutron activation analysis in different angles such as: basis principles, derivation of several equations, techniques, procedures, etc. In addition, we have associated in each part of general work, some examples of our technical developments and applications of NAA method in several fields. All analytical works were executed in our centre using irradiation facilities of Es-Salam research reactor and all necessary equipments installed in the NAA department to cover all steps of analytical process.

ACKNOWLEDGMENTS

Thanks are due to Dr Derdour Mohamed the responsible of COMENA and Mr Kerris Abdelmoumen General Director of CRNB for financial support. Grateful acknowledgment is made to Mr Salhi Mhamed the director of techniques and nuclear applications division for his highly valuable assistance. Special thanks are due to all colleagues involved for their help during fifteen years.

REFERENCES

1. Greenberg, R.R., et al., Spectrochimica Acta Part B, 66 issues 3-4; 2011, p193-241.
2. Bé, M. M., Chisté. V., BNM-LNHB/CEA- Tables des radionucléides, 01/12/01-6/4/2004 ; 2004.
3. De Corte, F.,The k0 Standardization Method: A Move to the Optimization of NAA, Gent University; 1987.
4. De Corte, F., et al.,J. Radioanal. Nucl. Chem., volume 197; 1994, p93.
5. Acharya, R., Chatt, A.,. J. Radioanal. Nucl. Chem., volume 257; 2003, p525.
6. Simonits, A., De Corte, F., Hoste, J., J. Radioanal. Chem., volume 24; 1975, p. 31.
7. Lin, X., Li, X., J. Radioanal. Nucl. Chem., volume 223; 1997, p. 47.
8. Moens, L., De Corte, F., Simonits, A., De Wilspelaere, A., Hoste, J., J. Radioanal. Chem., volume 52 ; 1979, p. 379.
9. De Corte, F., Simonits, A., De Wilspelaere, A., Hoste, J., J. Radioanal. Nucl. Chem., volume 113; 1987, p145.

10. Jovanovic, S., De Corte, F., Moens, L., Simonits, A., Hoste, J., J. Radioanal. Nucl. Chem., volume 82; 1984, p.379.

11. Alghem, L., Ramdhane ,M., S. Khaled, T. Akhal., The development and application of k0-standardization method of neutron activation analysis at Es-Salam research reactor, Nucl. Instr. and methods., 556; 2006, p386–390.

12. Mustra, C. O., Freitas, M. C., Alimeida, S. M. (). J. Radioanal. Nucl. Chem. volume 257; 2003, p539.

13. HyperLab 2005 System: Installation and Quick Start Guides. HyperLabs Software, Budapest, Hungary; 2005.

14. De Corte, F, Simonits, A, De Wispelaere, A., Hoste, J., Accuracy and applicability of the k0-standardozation method. J Radioanal Nucl Chem volume 113; 1987,. p145–154.

15. Alghem Hamidatou, L., Ramdhane, M., Characterization of neutron spectrum at Es- Salam Research Reactor using Høgdahl convention and Westcott formalism for the k0-based neutron activation analysis, Journal of Radioanalytical and Nuclear volume 278 issue (3); 2008, p627-630.

16. Dung, H.M., Hien, P.D., J. Radioanal. Chem. volume 257; 2003, p643.

17. De Corte, F., Simonits, A., Bellemanns, F, Freitas, MC., Jovanovic, S, Smodis, B., Erdtmann G., Petri, H., De Wispelaere, A., Recent advances in the k0-standardization of neutron activation analysis: Extensions, applications, prospects. J Radioanal Nucl Chem volume 169; 1993, p125–158.

18. Kennedy, G., ST-Pierre, J., Wang, K., Zhang, Y., Preston, J., Grant, C., Vutchkov, M., Activation constants for Slowpoke and MNS Reactors calculated from the neutron spectrum and k0 and Q0 values. J Radioanal Nucl Chem volume 245; 2000, p167–172.

19. Acharya, RN., Nair, AGC., Reddy, AVR., Manohar, SB. () Validation of a neutron activation analysis method using k0-standardization. Appl. Radiation Isotopes, volume 57; 2002, p391–398.

20. Kragten, J., Calculating standard deviations and confidence intervals with a universally applicable spreadsheet technique, Analyst volume 119; 1994, p2161–2166.

21. Rees, C.E., Error propagation calculations, Geochim. Cosmochim. Acta volume 48; 1984

22. Ellison, S.L.R., Rosslein, M., Williams, A., Quantifying Uncertainty in Analytical Measurements (2nd ed.; QUAM:2000.P1), Eurachem/CITAC; 2000.

23. Robouch, P., Arana, G., Eguskiza, M., Pommé, S., Etxebarria, N., Uncertainty budget for k0-NAA, J. Radioanal. Nucl. Chem. volume 245; 2000, p195–197.

24. Joint Committee for Guides in Metrology, Evaluation of measurement data - Supplement 1 to the "Guide to the expression of uncertainty in measurement" - Propagation of distributions using a Monte Carlo method JCGM101:2008 (ISO/IEC Guide 98-3-1); 2008.

25. Joint Committee for Guides in Metrology, Evaluation of measurement data — Guide to the expression of uncertainty in measurement (2008) JCGM 101:2008 (ISO/IEC Guide 98-3); 2008.

26. Bode, P., Instrumental and organizational aspects of a neutron activation analysis laboratory Ph.D. dissertation, Delft University of Technology, Delft; 1996.

27. IAEA TEC-DOC, 1215; 2001.

28. Rossbach, M., Emons, H., Groemping, A., Ostapczuk, P., Schladit, J.D., Quality control strategies at the environmental specimen bank of the Federal Republic of Germany, Harmonization of Health-Related Environmental Measurements Using Nuclear and Isotopic Techniques, (Proc. Symp. Hyderabad, India, 1996), IAEA, Vienna; 1997, p89–100.

29. International Atomic Energy Agency, Applications of Isotopes and Radiation in Conservation of the Environment (Proc. Symp. Karlsruhe, Germany, 1992), IAEA, Vienna; 1992.

30. International Atomic Energy Agency, Harmonization of Health-Related Environmental Measurements Using Nuclear and Isotopic Techniques, (Proc. Symp. Hyderabad, India, 1996), IAEA, Vienna; 1997.

31. Zeisler, R., Guinn, V.P., Nuclear Analytical Methods in the Life Sciences, Biol. Trace El. Res., 26/27, Humana Press, Clifton, New Jersey; 1990.

32. Kucera, J., Obrusnik, I., Sabbioni, E., Nuclear Analytical Methods in the Life Sciences, Biol. Trace El. Res., 43/45, Humana Press, Totowa, New Jersey; 1994.

33. Chai, Z., (Ed.), Nuclear Analytical Methods in the Life Sciences, (1999). Biol. Trace El. Res.71, Humana Press, Totowa, New Jersey.

34. Ehmann, W.D., Yates, S.W.Nuclear and radiochemical analysis, Anal. Chem. 58 49R-65R, 1986.

35. Ehmann, W.D., Yates, S.W., Nuclear and radiochemical analysis, Anal. Chem. 60 42R–62R; 1988.

36. Ehmann, W.D., Robertson, J.D., Yates, S.W., Nuclear and radiochemical analysis, Anal. Chem. 62 5OR–70R; 1990.

37. Ehmann, W.D., Robertson, J.D., Yates, S.W., Nuclear and radiochemical analysis, Anal. Chem. 64 1R–22R; 1992.

38. Ehmann, W.D., Robertson, J.D., Yates, S.W., Nuclear and radiochemical analysis, Anal. Chem. 66 229R–251R; 1994.

39. Application reviews, Analytical Chemistry 63, No. 12 1R-324R; 1991.

40. Application reviews, Analytical Chemistry 65, No. 12 1R-484R; 1993.

41. Application reviews, Analytical Chemistry 67, No. 12 1R-583R; 1995.

42. Application reviews, Analytical Chemistry 69, No. 12 1R-328R; 1997.

43. Khaled, S., Mouzai, M., Ararem, A., Hamidatou, L., Zouranen, B., Elemental analysis of traditional medicinal seeds by instrumental neutron activation analysis. J Radioanal Nucl Chem volume 281; 2009, p87–90.

44. Hamidatou, L. A,. Khaled, S., Akhal, T., Ramdhane, M., Determination of trace elements in cigarette tobacco with the k0-based NAA method using Es-Salam research reactor. J Radioanal Nucl Chem, volume 281; 2009, p535–540

45. Gülovali, M.C., Gunduz, G., J. Radioanal. Chem., volume 78; 1983, p189-198.

46. Abedinzadeh, Z., Razechi, M., Parsa, B., J. Radioanal. Chem. volume 35, 1977,p373-379.

47. Vega-Carrillo, H.R., Iskander, F.Y., Manzanres-Acuna, E., J. Radioanal. Nucl. Chem. Lett., volume 200; 1995, p137-145.

48. Munita, C.S., Mazzillil, B.P., J. Radioanal. Nucl. Chem. Lett, volume 108; 1986, p217-227.

49. Hamidatou, L. A., Khaled, S., Mouzai, M., Zouranen, B., Ararem, A., Alghem, A., Ramdhane, M., Instrumental neutron activation analysis of milk samples using the k0-standardization method at Es-Salam research reactor. Phys. Chem. News volume 45; 2009, p44-47.

50. Akhal, T., Mouzai, M., Sana, Y., Ladjel, M., Application of RNAA to the proportioning of iodine in food salt. Phys. Chem. News volume 45; 2009, p34-38.

51. Hamidatou, L. A., S. Dakar, S. Boukari, k0-NAA quality assessment in an Algerian laboratory by analysis of SMELS and four IAEA reference materials using Es-Salam Research reactor, Journal of Nuclear Instruments and methods in physics research section A., 682 (2012) p 75–78.

52. Hamidatou, L., Benkharfia, H Experimental and MCNP calculations of neutron flux parameters in irradiation channel at Es-Salam reactor, J. Radioanal. Nucl. Chem., V287, N° 3 (2011) p971-975.

2

WATER DYNAMICS IN SHEWANELLA ONEIDENSIS AT AMBIENT AND HIGH PRESSURE USING QUASI-ELASTIC NEUTRON SCATTERING

Fabrizia Foglia[1], Rachael Hazael[1], Giovanna G. Simeoni, Marie-Sousai Appavou, Martine Moulin, Michael Haertlein, V.Trevor Forsyth, Tilo Seydel, Isabelle Daniel, Filip Meersman & Paul F. McMillan[1]

[1]Chemistry Department, Christopher Ingold Laboratories, University College London, 20 Gordon Street, London WC1H 0AJ, UK.

ABSTRACT

Quasielastic neutron scattering (QENS) is an ideal technique for studying water transport and relaxation dynamics at pico- to nanosecond timescales and at length scales relevant to cellular dimensions. Studies of high pressure dynamic effects in live organisms are needed to understand Earth's deep biosphere and biotechnology applications. Here we applied QENS to study water transport in*Shewanella oneidensis* at ambient (0.1 MPa) and high (200 MPa) pressure using H/D isotopic contrast experiments for normal and perdeuterated bacteria and buffer solutions to distinguish intracellular and transmembrane processes. The results indicate that intracellular water dynamics are comparable with bulk diffusion rates in aqueous fluids at ambient conditions but a significant reduction occurs in high pressure mobility. We interpret this as due to enhanced interactions with macromolecules in the nanoconfined environment. Overall diffusion rates across the cell envelope also occur at similar rates but unexpected narrowing of the QENS signal appears between momentum transfer values $Q = 0.7–1.1$ Å^{-1} corresponding to real space dimensions of 6–9 Å. The relaxation time increase can be explained by correlated dynamics of molecules passing through Aquaporin water transport complexes located within the inner or outer membrane structures.

INTRODUCTION

Water dynamics inside cells and across the cell envelope are key to biological processes[1,2,3,4]. Although it is thought that water mobility must be significantly slowed by macromolecular interactions within the crowded intracellular environment, results from NMR and quasi-elastic neutron scattering (QENS) studies have revealed diffusion rates that more closely resemble bulk aqueous solutions[5,6]. QENS is a powerful technique for studying correlated proton dynamics and relaxation processes over a wide range of momentum (Q) and energy ($\hbar\omega$) transfers at nano- to picosecond timescales relevant to cellular biology[7,8,9]. In our experiment these correspond to a spectral broadening of the elastic line exceeding the energy resolution by a few tenths of meV. Therefore, the dynamics probed by QENS are readily distinguished from vibrational excitations or events such as enzyme catalysis that occur on significantly faster (fs) or slower (ms) timescales, respectively. QENS signals arise from incoherent scattering processes. Because of the large differences in H/D neutron scattering cross sections, isotopic contrast experiments can be designed to focus on the dynamics occurring within specific regions of the biological cell, including the intracellular medium (Im) or across membrane structures. QENS has already been developed to study high pressure effects on dynamic processes both in aqueous media and for biologically important macromolecules[10,11]. Here we extended the application of the QENS technique to investigate water dynamics in H/D substituted samples of live *Shewanella oneidensis* bacteria at ambient (0.1 MPa) and high (200 MPa) pressures relevant to subsurface biological processes.

Observations of live bacteria at >10 km below the ocean surface as well as samples recovered from continental and suboceanic drilling projects have revealed microbial lifeforms existing at up to at least P = 110 MPa[12,13]. Understanding the behavior of microorganisms exposed to HP conditions is also necessary to optimize food preservation and other biotechnologies where pressures in the 200–700 MPa range are applied to eliminate harmful bacteria[14]. We selected *S. oneidensis* as a model for our study. This prokaryotic organism exists under both aerobic and anaerobic conditions and its genus contains several piezophilic ("pressure-loving") species[15,16,17]. Its metal reduction (MR) capacities are being harnessed for environmental remediation and fuel cell applications[18,19], and its metabolic processes have been studied by *in situ*spectroscopy under HP conditions[20,21,22]. *S. oneidensis* has also been investigated to study its pressure adaptation and survival under conditions extending into the GigaPascal (P > 1 GPa) range[20,23].

RESULTS AND DISCUSSION

Our QENS experiments were carried out at the time-of-flight spectrometer TOFTOF (Garching, Germany)[24] that allowed recording with low background levels necessary for successful analysis of the H/D isotopic contrast experiments, over a wide *Q* range (0.2–1.8 Å^{-1}) corresponding to real space correlation lengths ~3–30 Å (Fig. 1). A detailed discussion of the experimental details, including sample preparation, instrumentation and data analysis protocols is presented given in the Materials and Methods section below.

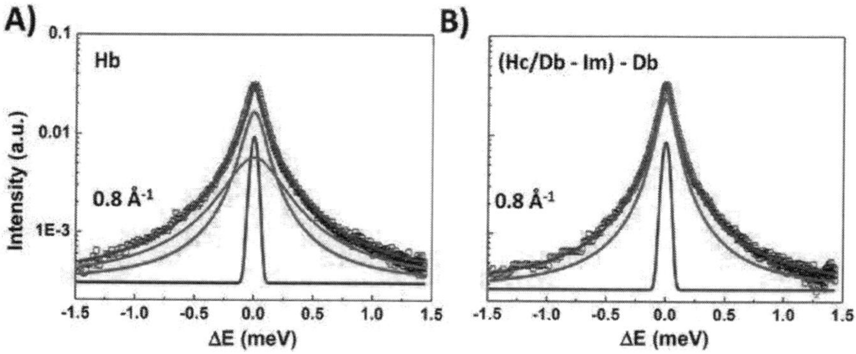

Figure 1. Analysis of the neutron dynamic scattering function $S(Q, \omega)$ at ambient pressure (0.1 MPa) and $Q = 0.8\,Å^{-1}$ for the hydrogenated buffer solution ((**A**): Hb) used for bacterial suspensions and an isotopic contrast ((**B**): [Hc/Db-Im]-Db; Im represents the intracellular medium) designed to highlight dynamics across the cell envelope. The data and lineshape fits are shown with a logarithmic intensity scale to highlight QENS contributions. The central Gaussian (blue) due to elastic scattering defines the instrumental resolution function. For aqueous solutions both translational (magenta) and rotational (green) Lorentzian components were used to fit the QENS lineshapes. These two components are related to τ_D (translational relaxation component corresponding to water diffusion) and τ_R (molecular rotational relaxational processes). The global fit (red continuous curve) is overlain on the data points (black). The isotopic contrast datasets (**B**) used to analyse transmembrane dynamics were fit using a single Lorentzian function. The energy transfer scale shown in meV is converted to reciprocal time (s^{-1}) units using $\Delta E = \hbar\omega$.

BULK AQUEOUS BUFFER SOLUTIONS

We tested our approach by examining water dynamics in the bulk aqueous buffer solutions (non isotopically substituted buffer Hb; perdeuterated (98%) buffer Db) that served as the bacterial growth and suspension media for our experiments. The main electrolyte species present are Na^+ and Cl^- with structure-breaking to borderline properties along with structure-making phosphate anions25. The QENS broadening arising from diffusive mechanisms can be described by a sum of Lorentzian functions, whose number and relative contribution depends on the specific processes involved. Focusing on the main narrow component associated with translational relaxation, fitting the full width at half maximum (FWHM) $\Gamma_T(Q^2)$ relations to a Singwi-Sjölander (*SS*) jump model[7] (see Methods section below) led to diffusion coefficients $D_T = 2.70_{\pm 0.20} \times 10^{-5}$ and $2.00_{\pm 0.11} \times 10^{-5}$ cm^2s^{-1} for Hb and Db, respectively, at ambient conditions (Fig. 2A,B, Table 1). These results are in excellent agreement with tabulated values for water and aqueous solutions at or near ambient temperature. At 200 MPa the diffusion coefficients are lowered by 9–11% for the aqueous electrolyte solutions. The reduced QENS linewidths observed for Db *vs* Hb liquids at both pressures occur due to de Gennes narrowing, combined with atomic mass effects on the molecular diffusion coefficients[7,11,26,27].

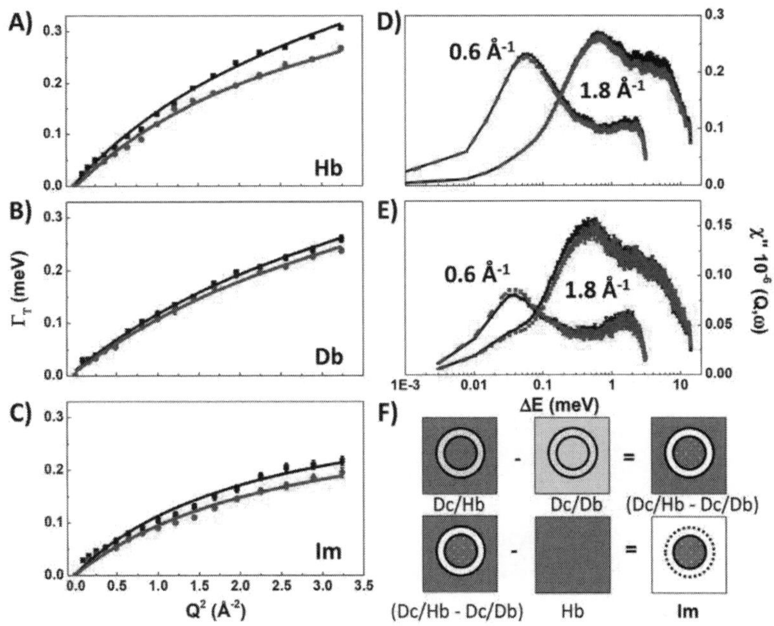

Figure 2. Water dynamics in the aqueous suspension (panel (**A**): Hb, H-buffer; panel (**B**): Db, D-buffer) and intracellular medium (panel (**C**): Im) from analysis of the QENS linewidth (Γ for the translational diffusion component) as a function of momentum transfer (Q^2). The isotopic contrast combination used to obtain Im was [(Dc/Hb–Dc/Db)–Hb] (panel (**F**)). At left are shown the $\Gamma_T(Q^2)$ dependencies at 0.1 (black) and 200MPa (red) fit to a *SS* jump model. At right are shown the $\chi'(Q, \omega)$ susceptibility functions determined at $Q = 0.6$ and 1.8 $Å^{-1}$ for the aqueous solutions.

Table 1. Diffusion parameters extracted from fits by applying the *SS* jump model to fit $\Gamma_T(Q^2)$ within $Q = 0.2–1.0$ and $0.2–1.8$ Å^{-1} ranges at 0.1 and 200 MPa.

Sample	Pressure/MPa	Diffusion parameters applying the SS jump model over Q-range indicated									
		Q-range: 0.2–1.0 Å^{-1}			Q-range: 0.2–1.8 Å^{-1}						
		$D_T \times 10^{-5}$/cm^{2}s^{-1}	τ_0/ps	2W	$D_T \times 10^{-5}$/cm^{2}s^{-1}	τ_0/ps	2W	ΔP/%	τ_0/ps	ΔP/%
Hydrogenated buffer (Hb)	0.1	—	—	—	2.70 ± 0.20	1.0 ± 0.10	0.01	−11	1.68 ± 0.06	+18
	200	—	—	—	2.40 ± 0.19	1.22 ± 0.13	0.01		2.05 ± 0.09	
Deuterated buffer (Db)	0.1	—	—	—	2.00 ± 0.11	1.02 ± 0.09	0.02	−9	1.80 ± 0.06	+10
	200	—	—	—	1.82 ± 0.20	1.15 ± 0.16	0.02		2.00 ± 0.09	
Intracellular medium (Im) (Dc/Hb–Dc/Db)–Hb	0.1	—	—	—	2.43 ± 0.14	1.81 ± 0.07	0.05	−18	1.85 ± 0.10	+26
	200	—	—	—	2.00 ± 0.12	2.00 ± 0.10	0.05		2.50 ± 0.09	
[(Hc/Db)–Db]	0.1	2.12 ± 0.30	3.18 ± 0.21	0.17	2.57 ± 0.40	2.04 ± 0.10	0.07	−18	1.92 ± 0.12	+12
	200	1.67 ± 0.40	3.71 ± 0.19	0.24	2.11 ± 0.30	2.11 ± 0.09	0.08		2.18 ± 0.10	
[(Dc/Hb)–Hb]	0.1	2.12 ± 0.40	3.22 ± 0.26	0.17	2.45 ± 0.36	1.97 ± 0.25	0.07	−18	1.95 ± 0.09	+11
	200	1.82 ± 0.30	3.46 ± 0.22	0.20	2.05 ± 0.30	2.13 ± 0.08	0.08		2.20 ± 0.09	
[((H/D)–Im)–Db]	0.1	1.75 ± 0.36	2.27 ± 0.10	0.09	2.43 ± 0.24	1.89 ± 0.09	0.04	−20	—	
	200	1.52 ± 0.38	2.45 ± 0.33	0.09	1.95 ± 0.30	2.15 ± 0.08	0.06		—	
[((D/H)–Im)–Hb]	0.1	1.82 ± 0.25	2.21 ± 0.30	0.10	2.64 ± 0.36	2.01 ± 0.17	0.05	−21	—	
	200	1.52 ± 0.36	2.49 ± 0.27	0.11	2.10 ± 0.21	2.35 ± 0.12	0.05		—	

As we investigated the $\Gamma_T(Q^2)$ properties of each dataset, we also carried out a complementary examination of the susceptibility function $\chi'(Q, \omega)$ normalized to the frequency (energy transfer) scale[28,29]. At finite temperature the intensity of any experimentally measured scattering function $S(Q, \omega)$ is asymmetric with respect to the energy loss (ω) axis due to the Bose-Einstein thermal population factor. Studying the form of $\chi'(Q, \omega)$ reveals the proper distribution of spectral weight among the different dynamic processes, along with additional useful information concerning the static structure factor $S(Q)$. Furthermore, using a stretched energy scale with respect to the corresponding $S(Q = \text{const}, \omega)$ spectrum allows a clear separation between and determination of the number of relaxation mechanisms contained within the density of states responsible for the diffusive dynamics. Similar interpretations can also be achieved by developing a sophisticated analysis of relationships between fitting parameters used to reproduce the QENS linewidth result; however, we preferred to crosscheck the fit analyses by using the $\chi'(Q, \omega)$ results as an independent method. For simplicity we show the details only for the case of the bulk aqueous solutions, even though its analysis has supported the interpretation of all the reported results and also provided an *in situ* marker of the hydrostatic conditions achieved during the experiments. In Fig. 2D,E we report the imaginary susceptibility of both Hb and Db at two Q values representative of long ($Q = 0.6$ Å^{-1}) and short ($Q = 1.8$ Å^{-1}) range dynamics. In the case of the Hb data (Fig. 2D), corresponding to a mostly incoherent signal, the two observed peaks have comparable intensity. However, in the case of Db (Fig. 2E), where the signal is dominated by coherent scattering, the intensity of the peaks is considerably modified by the influence of the static structure factor S(*Q*).

INTRACELLULAR MEDIUM (IM)

We next examined water dynamics in the intracellular medium (Im) using isotopic contrasts derived from QENS datasets for: (i) unlabelled bacterial cells (Hc) suspended in deuterated buffer solution (Hc/Db); (ii) perdeuterated cells (Dc) suspended in deuterated buffer (Dc/Db) and (iii) perdeuterated cells in unlabelled buffer (Dc/Hb), to construct the isotopic contrast [(Dc/Hb–Dc/Db)–Hb] (Fig. 2, Table 1). Fitting the $\Gamma_t(Q^2)$ data by the SS jump model led to an intracellular diffusion coefficient $D_T = 2.43_{\pm 0.14} \times 10^{-5}$ cm^{2}s^{-1}, intermediate between the values for Hb and Db buffer solutions. This result indicates that water diffusion in the intracellular medium is indistinguishable from the bulk over the examined Q-range corresponding to correlated motions extending between approximately 3–30 Å in real space. Jasnin *et al.* likewise observed no differences in *E. coli* cytoplasm dynamics from diffusion in bulk aqueous media examined by QENS at ambient pressure[6,7,8,9]. Similar experiments on hydrogels with water confined within nanoscopic pores show that water diffusion over these length scales is essentially the same as that in bulk fluids[30]. As pressure was increased to 200 MPa we observed that the cytoplasmic mobility in *S. oneidensis* became reduced by 18%, nearly twice that observed for the Hb or Db aqueous buffer solutions (−11%, −9%) (Fig. 2C, Table 1). The pressure effect on the $\Gamma_T(Q^2)$ is already obvious from initial inspection of the QENS data sets that clearly show an enhanced signal at large *Q* values for the Im compared with bulk Hb, Db buffer solutions (Fig. 3). This dynamic slowdown is accompanied by a 26% increase in the rotational relaxation time τ_R, compared with only 10–18% for the bulk aqueous electrolyte solutions (Table 1). We suggest that these effects indicate enhanced interactions between H_2O and intracellular macromolecules becoming more dominant as the volume within the cell is reduced at high pressure.

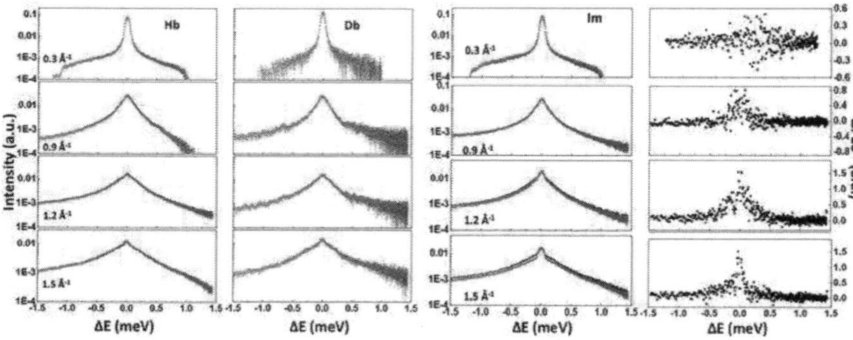

Figure 3: Selected QENS data at different *Q* values and 0.1 *vs* 200 MPa pressures. Left panel: Comparison between neutron dynamic scattering functions S(*Q*, ω) at 0.1 (black symbols) and 200 MPa (red symbols) for aqueous buffer solutions (H-buffer, Hb and D-buffer, Db) at $Q = 0.3$, 0.9, 1.2 and 1.5 $Å^{-1}$. Right panel: Comparison between neutron dynamic scattering functions S(*Q*, ω) at 0.1 (black symbols) and 200 MPa (red symbols) for water dynamics within the intracellular medium (Im, obtained from the following subtraction between isotopic contrasts: [(Dc/Hb–Dc/Db)–Hb]) at $Q = 0.3$, 0.9, 1.2 and 1.5 $Å^{-1}$, showing the development of a significant pressure effect on the QENS profile at high *Q* values. This is absent in the case of Hb and Db buffer solutions. At far right are shown the corresponding subtractions between Im scattering profiles obtained at 0.1 and 200 MPa to highlight the effect of pressure on the QENS profile.

TRANSPORT ACROSS THE CELLULAR ENVELOPE

To investigate water transport across the cell envelope we first examined the isotopic contrasts (Hc/Db–Db) and (Dc/Hb–Hb) assuming that water diffusion rates inside the cell remained the same as in bulk aqueous media at high pressure. In a second refined version of the model we then constructed the isotopic contrasts [((H/D)–Im)–Db] and [((D/H)–Im)–Hb] that take into account the differential effect on Im diffusion at pressure compared with the Hb and Db buffer solutions noted above (Fig. 4). Both approaches gave similar results.

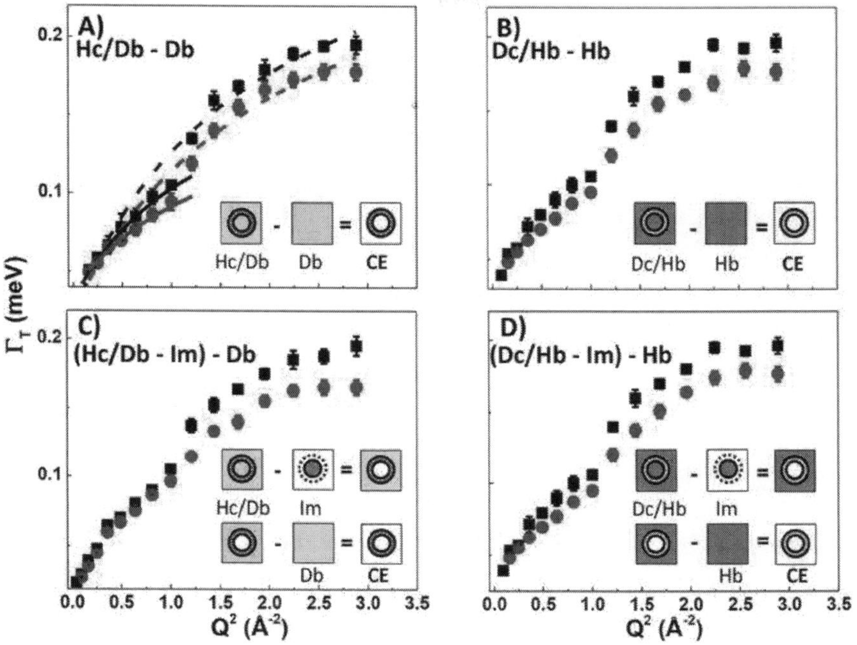

Figure 4: Water dynamics across the cell envelope (CE) at 0.1 (black) and 200MPa (red), visualised using different isotopic contrast subtractions. (**A**) (Hc/Db – Db), with *SS* jump model fits to the overall cell envelope (0.2–1.8 Å^{-1}) shown as dashed lines and restricted to the 0.2–1.0 Å^{-1} range (solid lines). (**B**) Data points for the corresponding isotopic contrast (Dc/Hb – Hb). Panels (**C**) and (**D**): isotopic contrasts [(Hc/Db – Im) – Db)] and [(Dc/Hb – Im) – Hb)] showing data points taking into account differential effects of pressure on Im compared with bulk media (Hb, Db).

The $\Gamma_T(Q^2)$ data present a similar picture of water dynamics to the bulk and Im media but an unexpected narrowing of the QENS linewidths occurs between $Q^2 = 0.5–1.2$ Å^{-2}. A small downwards deviation in $\Gamma(Q^2)$ does appear to be present for the Im data set (Fig. 2) between $Q^2 = 1–2.3$ Å^{-2}, however, that is much less pronounced than for the isotopic contrast datasets used to analyze the transmembrane behavior. We note that the Im dataset subtractions used to study the intracellular dynamics (Im) might have been slightly affected by transmembrane water diffusion processes, but to a much smaller extent than observed for the datasets presented in Fig. 4 and discussed in this section.

Fitting the SS model to the ensemble of data points at long ($Q < 0.7$ Å^{-1}) and short ($Q > 1.1$ Å^{-1}) correlation lengths we obtained $D_T = 2.43$–2.64×10^{-5} and 1.95–2.10×10^{-5} cm^{2}s^{-1} (at 0.1 and 200 MPa, respectively), depending on the isotopic contrast model used (dashed line fit illustrated in Fig. 4, panel A). These values represent the overall diffusion coefficients throughout the entire cell envelope. Both they and their pressure dependence are nearly identical to those for diffusion within the Im (Table 1). However, we note that if only data from longer correlation lengths ($Q < 1$ Å^{-1}) had been considered in the QENS analysis (solid line in Fig. 4A) the result would indicate significantly lowered apparent transmembrane diffusion coefficients ranging between 1.75–1.82×10^{-5} and 1.52–1.72×10^{-5} cm^{2}s^{-1} (i.e., at 0.1 and 200 MPa, respectively) (Fig. 4, Table 1).

We now examine the implications of these results and provide an initial interpretation of the unusual form of the $\Gamma_T(Q^2)$ relations. An initial downwards deviation of the data from the initial progression of the curve near 0.7 Å^{-1} is followed by a rapid upturn as the data regains the "normal" diffusion relation at approximately 1.1 Å^{-1}. In an early analysis of the data, we considered the possibility that this might be associated with two superposed motions (i.e., transmembrane diffusion *vs* nanoconfined water dynamics within the membrane structures). However, it quickly became clear that the jump model, or any superposition of jump models, could not describe the water dynamics in any meaningful or justifiable way in terms of likely relaxation or nanoconfinement scenarios. However, a simpler model can be advanced to interpret the transmembrane $\Gamma_T(Q^2)$ relations by examining the relaxation timescales implied by the QENS data, along with models that are becoming established for molecular water transport across the cell membrane structures *via* Aquaporin protein complex channels.

First we note that the unexpected slope changes can be interpreted as a "dip" relative to the $\Gamma_T(Q^2)$ relation expected for a jump diffusion model applied to a single diffusion process, between $Q = 0.7$–1.1 Å^{-1}. Each data point within this range represents an unexpected narrowing of the QENS linewidth and hence a correspondingly enhanced relaxation time for those correlated water transport motions occurring with correlation lengths between 6–9 Å in real space (Fig. 4). This spatial dimension is identified with that of the central region of Aquaporin (Aqp) channels that are already known to transport water molecules across the cell membrane in animal and bacterial cells[31,32,33,34]. AqpZ is known to be present in *Shewanella*. We note that all four isotopic contrast combinations shown in Fig. 4 show the same result. If the anomaly in the $\Gamma_T(Q^2)$ datasets had been due to a systematic instrumental artefact, this would have been removed by the subtraction procedures. However, if such an artefact had been present in only one of the datasets, this would have given rise to anomalous slope changes occurring in opposite directions for different contrast combinations. The data inFig. 4 show that this is clearly not the case, and we can examine the physical interpretation of the QENS line narrowing effect.

MD simulations combined with X-ray crystallography have previously shown that approximately 7–9 H_2O molecules traverse the Aqp channel in single file in a spatially correlated motion, with a reversal of the molecular orientation occurring midway through the channel that ensures transport *via* intact molecular units[34]. Invoking such highly correlated H_2O transport dynamics can then explain the unusually long lifetime recorded by our QENS data, that correspond to 6.3 ps at 1 Å^{-1}, compared with 5.3 ps expected by interpolating between the bulk diffusion dynamics occurring at longer and shorter Q values (Fig. 3). Our QENS isotopic contrast experiments thus provide support for the model of highly correlated water transport dynamics occurring within transmembrane Aqp channels of *S. oneidensis*. The QENS linewidth exhibits a further reduction at 200 MPa indicating an ~7% increase in correlation time for water molecules in the transport channel as the applied pressure is increased (Fig. 4). This observation will have implications for the way in which water is transported across the cell wall for organisms subjected to high pressure conditions.

CONCLUSION

We used QENS at ambient (0.1 MPa) and high pressure (200 MPa) combined with a range of H/D contrast experiments on isotopically labelled samples to reveal details of water dynamics occurring within the intracellular medium and across the cell membrane for *Shewanella oneidensis* bacteria suspended in their growth/buffer media. The QENS linewidth data were analyzed according to molecular jump models to examine the effects of pressure on the water transport processes. At ambient pressure the intracellular (Im) mobility is indistinguishable from bulk water dynamics over correlation lengths extending between 3–30 Å. However, the reduction in diffusion coefficients with pressure is significantly greater than for the bulk aqueous solutions indicating the presence of interactions with macromolecules within the nanoconfined cellular environment. The overall water diffusion rate across the cell envelope is similar to that in the intracellular medium at ambient and high pressure. However, an unusual narrowing of the QENS line occurs for $Q = 0.7–1.1$ Å^{-1} corresponding to an enhanced relaxation time for correlated water transport over a real space distance ~6–9 Å. This corresponds to H_2O molecules passing through the transmembrane Aqp channels embedded within the outer or inner cell membrane structures. Our HP-QENS studies do not reveal remarkable effects on H_2O diffusion either inside the cells or across the cell membrane at pressures up to 200 MPa, that are relevant to the range in which subsurface microbial life has been shown to exist. *In situ* studies of both aerobic and anaerobic metabolic processes enabled by monitoring the rates of formate oxidation and Se^{4+} or Fe^{3+}reduction have demonstrated a >3-fold decrease in respiration rates within the same pressure range20,21,22. Our QENS results reveal that this metabolic slowdown is not directly coupled to the intracellular or transmembrane water diffusion rates at least on the molecular length and time scales probed here.

MATERIALS AND METHODS

Bacterial sample preparation and isotopic substitutions

S. oneidensis samples (CIP 106686) obtained from the Collection Institut Pasteur (Paris, France) were grown at 30 °C in LB broth at 180 rpm after selecting a single colony from a LB agar plate and harvested at stationary phase (10^8 cells/mL). A perdeuterated strain was grown at the Deuteration Laboratory (ILL-PSB, Grenoble, France). The D-silantes medium used to grow and deuterate the bacteria prior to the neutron scattering experiments led to >98% labeling35. Bacterial samples were prepared at a concentration of 50 mg/mL and data recorded with acquisition time 6 h for each pressure point. The pressure ramp used to reach the target value was 20 MPa/min. Complementary laboratory experiments carried out using similar protocols to previous work23 showed a 14% bacterial survival rate following pressurization at 200 MPa for 6 h.

Because H or D incorporated within the macromolecular cellular structures are primarily non-labile36, our isotope contrast experiments allowed us to focus attention on mobile water species diffusing over correlated length scales up to 30–40Å, that are relevant to intracellular and trans-membrane dimensions. Proton exchange reactions and vibrational excitations within these systems occur on either much slower (ms) or faster (fs) timescales that were not imaged by our QENS experiments that are restricted to the picosecond (ps) regime.

The dataset subtractions made to construct the isotopic contrasts were enabled by the large difference in incoherent scattering cross section between H and D (2.05 and 80.27 barn, respectively), taking account of the fact that D is almost invisible relative to H when both are present. The D-labelled growth/suspension medium was not 100% deuterated and the resulting perdeuterated cells also retained a small quantity of hydrogenated component. Because of the scattering cross sections involved, the isotopic contrast Dc/Db could be used to remove most of the coherent scattering contribution that is mainly present in the elastic line, as well as incoherent scattering from the H-component that is essentially located in the inelastic part of the scattering pattern, so that the results we report are truly QENS data sets. The presence of the isotopic impurity also helped ensure that an incoherent signal was observed in these Dc/Db experiments.

Some D/H exchange occurs naturally *via* exchangeable protons when D-cells are placed in H-buffer, and *vice versa*. Each sample was resuspended three times in the relevant isotopic medium and left to equilibrate for at least 30 minutes before any data acquisition in order to ensure complete re-equilibration among any chemically exchangeable protons and deuterons in the bacterial sample37,38.

QENS experiments and data treatment protocols

QENS experiments were carried out at the time-of-flight spectrometer TOFTOF (Garching, Germany)24 that allowed recording with low background levels necessary for successful analysis of the H/D isotopic contrast experiments, over a wide (0.2–1.8 Å^{-1}) Q range. TOFTOF is a direct geometry spectrometer: the Qresolution is ensured by the large solid angle covered by the detector area, whereas the energy of the scattered neutrons is measured after the time-of-flight for travelling from the sample to the detectors. The instrumental energy resolution (0.06 meV) defined by the incident neutron wavelength ($\lambda = 6$ Å) and chopper frequency (12000 rpm) produced a constant Gaussian lineshape that was perfectly suited to the characterization of the QENS signals of interest.

Our experiments that present a new approach using neutron scattering techniques to investigate important biological and biophysical dynamics required about 18 months of dedicated technical and conceptual development. It is important to evaluate the reproducibility and reliability of our findings. We repeated the experiments twice over a period of several months to allow an assessment of all potential sources of uncertainty ranging from sample growth, deuteration and resuspension protocols, through possible changes in samples occurring during shipping and handling, to instrumental issues associated with neutron flux, humidity accumulation on detectors, cell alignment, etc., and testing of software and protocols for data handling. The two independent datasets provided fully consistent results, and they could be averaged to provide error bars that we associated with each experimental data point. These error bars were so minor that they appear included within the size of the data points shown in Fig. 2, for example. This gives us confidence in the physical interpretation of our results. However, we note that similar experiments should be carried out at other neutron scattering facilities to fully demonstrate the power of this valuable technique to give new information on the dynamics of water and biological macromolecules in live organisms potentially as they are exposed to extreme environmental conditions.

The neutron scattering data were handled and treated using the Large Array Manipulation Program (LAMP)39. This software package provided the standard routines for the proper calibration and normalization of TOF data files and converted the matrix I(2Θ, t) into the dynamic structure factor S(Q, ω) of interest. Data analysis was performed on spectra at different constant Q values. Further details are given below. HP-QENS data were acquired using an Al-body cell with an internal gauge used to continuously monitor the pressure. In contrast to many high pressure investigations that are affected by a huge background from the sample environment, the measurement of a liquid solution in such an Al-cell represents a favorable condition for QENS studies. At an incident wavelength of 6 Å the pressure cell is almost transparent: i.e., it exhibits 97% neutron transmission overall, no incoherent scattering in the QENS region, Bragg peaks outside the accessible Q-range, and phonons well separated in energy from any liquid features. Moreover, we chose a scattering geometry with the high pressure cell rotated by 45° clockwise with respect to the propagation axis of the incident neutron beam, thus allowing a continuous signal acquisition over the entire detector area (i.e., with no shaded detector area). Further details concerning the high pressure cell and the scattering geometry are provided in11.

Raw data files were normalized to the neutron flux and the acquisition time, thus removing any dependence on the neutron source and the experimental duration. The global detector efficiency was calibrated to the energy of the incoming neutron beam. Further detector-specific efficiency fluctuations were corrected individually after comparison with the purely incoherent signal scattered from an elemental vanadium (V) reference sample. This constituted a 1.1 mm thick flat plate that also provided a calibrant for the experimental elastic resolution function. Indeed to guarantee a uniform normalization over all spectra contained within a workspace dataset, the time-resolved scattered intensity recorded by each detector was normalized to the integrated elastic intensity scattered by the vanadium reference standard at the same 2Θ angular position.We note that in the case of direct geometry TOF spectrometers, the Qresolution is provided by the distinct angular coordinate 2Θ assigned to each detector. Time-of-flight and angular units were then converted into the energy and momentum transfer scale according to standard LAMP routines, developed to provide the exact transformation of the double differential cross section into the corresponding dynamic structure factor S(Q, ω). The intensity data from scattered neutrons were either grouped according to a regular (Q, E) grid (linear binning) to provide information on $S(Q, \omega)$, or in constant Q spectra by grouping the signal from neighboring detectors. Additional normalization steps and data analysis were carried out directly on the $S(Q = \text{const}, \omega)$ spectra using Microcal OriginPro 8.6. These normalizations considered A) the total quantity of material present in the beam and B) the correct quantity of buffer to be subtracted. With respect to point B), we had to take into account the volume fraction ($\phi = 5\%$) occupied by the bacterial cells. For step A) we note that the scattered intensity $I(2\Theta, t)$, and therefore $S(Q, \omega)$, is proportional to the amount of material in the beam and is already weighted for the corresponding coherent/incoherent cross section. Each set of $S(Q = \text{const}, \omega)$ spectra generated by the same $S(Q, \omega)$ matrix was normalized to the integral of $S(Q, \omega)$ over the entire Q range and the quasielastic ω range. The resolution of the time-of-flight experiment excluded any phononic contribution from the empty can.

The spectra were analyzed using a built-in least squares algorithm that takes into account a central Gaussian feature (to account for the energy resolution) along with up to three Lorentzian functions. The elastic component was removed to reveal the quasi-elastic line broadening of interest for each dataset as a function of the scattering vector Q.

Following data acquisition and discrimination of the QENS signal from the pure elastic, Gaussian scattering component determined by the instrument resolution function we proceeded with construction of the isotopic contrasts and comparison between 0.1 and 200 MPa profiles, followed by data fitting using a hierarchy of models developed to describe the relaxation dynamics and diffusion processes[6,7,8,9,10,11,40]. For studies of the bulk aqueous buffer media (Hb, Db) and the intracellular medium (Im) the data were best fit using two Lorentzian lineshapes representing translational relaxation corresponding to diffusional processes, and a rotational relaxation component. The QENS results across the cell membrane were best fit using only a single Lorentzian contribution related to the transmembrane diffusion (Fig. 1; Table 1). Each Lorentzian lineshape is associated with a half width at half maximum (Γ in meV) that reveals itself as a measure of the relaxation time for an excitation with correlation length Q, once it is converted into inverse timescale units

(s^{-1})*via* $\Delta E = \hbar\omega$. The translational linewidth (Γ_T) is analyzed as a function of *Q*using different models for the transport process[6,7,8,40].

The following protocols were applied to establish S(*Q*, ω) for the different cell/buffer combinations:

(i) S(*Q*, ω) [bacterial cells in buffer] = S(*Q*, ω) [bacterial cells in buffer within the Al HP container] – S(*Q*, ω) [empty container]

where cells (Hc, Dc) and aqueous buffer media (Db, Hb) were hydrogenated (H) or perdeuterated (D) varieties. A similar subtraction procedure was used for the buffer solutions alone:

(ii) S(*Q*, ω) [buffer] = S(*Q*, ω) [buffer within the Al HP container] – S(*Q*, ω) [empty container]

We initially applied a correction for self absorption by the empty high P Al container but this was neglected in later treatments as it made no difference to the results.

(iii) S(*Q*, ω) [bacterial cells] = S(*Q*, ω) [bacterial cells in buffer] – α × S(*Q*, ω) [buffer]

Here the factor $\alpha = (1-\phi)$ accounts for the macromolecular volume fraction (5%) occupied by the bacterial cells in solution. In contrast to macromolecules such as proteins, with a composition that can be determined readily by sequencing and other assay techniques, the exact molecular composition of the live cells is not known precisely. However, because the buffer does not produce in any intense peak in the elastic region, the elastic scattering from the cells gives a reliable estimate of the macromolecular concentration. We thus estimated ϕ and α by normalizing the elastic intensity of (Dc/Db) to that of (Dc/Hb) at the first diffraction maximum (1.8 $Å^{-1}$), where pure coherent scattering was expected to occur.

Dataset analyses

The relationship

$$\Gamma_T = D_T Q^2 \qquad (1)$$

that introduces the diffusion coefficient D_T (cm^2s^{-1}) applies when interactions between molecules are weak, and in general as $D_T \rightarrow 0$. Typically plots of Γ_T versus Q^2 are analyzed to provide estimates of D_T that are generally found to be comparable with NMR and bulk transport measurements[6,7,8,40].

As intramolecular interactions become significant the diffusion processes deviate from Fick's law and the dynamics are described using two (or more) characteristic timescales. A typical approach involves the use of molecular jump models between sites separated by average length, l, and mean site residence time, τ_0, during which the molecules undergo oscillatory motions:

$$\Gamma_T = \frac{D_T Q^2}{D_T Q^2 \tau_0 + 1} \qquad (2)$$

$$D_T = l^2/\tau_0 \qquad (3)$$

If the jump lengths are distributed randomly, the diffusion is usefully described by a Singwi-Sjölander (*SS*) jump model, in which each water molecule executes oscillations during a mean residence time (τ_0) then diffuses during time τ_1 to a second site[6,7,8,9,10,11,26]. Typically, $\tau_0 \gg \tau_1$ and the fitting model then reduces to:

$$\Gamma_T = \frac{1}{\tau_0}\left[1 - \frac{\exp(-2W)}{D_T Q^2 \tau_0 + 1}\right] \qquad (4)$$

2W is a Debye-Waller factor $(2W = (1/6)Q^2\langle R^2\rangle)$ where $\langle R^2\rangle$ is the mean square displacement in the oscillatory part of the motion. Here we present our results using the *SS* model to analyze the QENS datasets, although other models were examined during the course of the study.

Additional Information

How to cite this article: Foglia, F. *et al.* Water Dynamics in *Shewanella oneidensis* at Ambient and High Pressure using Quasi-Elastic Neutron Scattering. *Sci. Rep.* **6**, 18862;

REFERENCES

1. Ball, P. Water as an active constituent in cell biology. Chem. Rev. 108, 74–108 (2008).

2. Clegg, J. S. Intracellular water and the cytomatrix: some methods of study and current views. J. Cell. Biol. 99, 167s–171s (1984).

3. Chaplin, M. Do we underestimate the importance of water in cell biology? Nature Rev. Mol. Cell Biol. 7, 861–866 (2006).

4. Mentré, P. An introduction to "water in the cell": tamed hydra? Cell. Mol. Biol. 47, 709–715 (2001).

5. Le Bihan, D. The 'wet mind': water and functional neuroimaging. Phys. Med. Biol. 52, R57–R90 (2007).

6. Jasnin, M., Stadler, A., Tehei, M. & Zaccai, G. Specific cellular water dynamics observed in vivo by neutron scattering and NMR. Phys. Chem. Chem. Phys. 12, 10154–10160 (2010).

7. Bée, M. Quasielastic Neutron Scattering: Principles and Applications in Solid State Chemistry, Biology and Material Science. (Adam Higer, Bristol, 1988).

8. Jasnin, M., Moulin, M., Haertlein, M., Zaccai, G. & Tehei, M. Down to atomic-scale intracellular water dynamics. EMBO Rep. 9, 543–547 (2008).

9. Frölich, A., Gabel, F., Jasnin, M., Lehnert, U. & Oesterhelt, D. From shell to cell: neutron scattering studies of biological water dynamics and coupling to activity. Faraday Disc. 141, 117–130 (2008).

10. Bove, L. E. et al. Translational and rotational diffusion in water in the Gigapascal range. Phys. Rev. Lett. 111, 185901 (2013).

11. Appavou, M. S., Busch, S., Doster, W., Gaspar, A. & Unruh, T. The influence of 2 kbar pressure on the global and internal dynamics of human hemoglobin observed by quasielastic neutron scattering. Eur. Biophys. J. 40, 705–714 (2011).

12. Picard, A. & Daniel, I. Pressure as an environmental parameter for microbial life: a review. Biophys. Chem. 183, 30–41 (2013).

13. Meersman, F. et al. High-pressure biochemistry and biophysics. Rev. Mineral. Geochem. 75, 607–648 (2013).

14. Aertsen, A., Meersman, F., Hendrickx, M. E. G., Vogel, R. F. &Michiels, C. W. Biotechnology under high pressure: applications and implications. Trends Biotechnol. 27, 434–441 (2009).

15. Kato, C. & Nogi, Y. Correlation between phylogenetic structure and function: examples from deep-sea Shewanella. FEMS Microbiol. Ecol.35, 223–230 (2001).

16. Fredrickson, J. K. & Romine, M. F. Genome-assisted analysis of dissimilatory metal-reducing bacteria. Current Opinion Biotechnol. 16, 269–274 (2005).

17. Heidelberg, J. F. et al. Genome sequence of the dissimilatory metal ion-reducing bacterium Shewanella oneidensis. Nature Biotechnol. 20, 1118–1123 (2002).

18. Ge, Z., Li, J., Xiao, L., Tong, Y. & He, Z. Recovery of electrical energy in microbial fuel cells. Env. Sci. Tech. Lett. 1, 137–141 (2013).

19. Logan, B. E. & Regan, J. M. Electricity-producing bacterial communities in microbial fuel cells. Trends Microbiol. 14, 512–518 (2006).

20. Sharma, A. et al. Microbial activity at gigapascal pressures. Science 295, 1514–1516 (2002).

21. Picard, A. et al. Monitoring microbial redox transformations of metals and metalloid elements under high pressure using in situ X-ray absorption spectroscopy. Geobiol. 9, 196–204 (2011).

22. Picard, A., Testemale, D., Hazemann, J. L. & Daniel, I. The influence of high hydrostatic pressure on bacterial dissimilatory iron reduction.Geochim. Cosmochim. Acta 88, 120–129 (2012).

23. Hazael. R. et al. Laboratory investigation of high pressure survival inShewanella oneidensis MR-1 into the gigapascal pressure range.Frontiers Microbiol. 5, 612–621 (2014).

24. Unruh, T., Neuhaus, E. & Petry, W. The high-resolution time-of-flight spectrometer TOFTOF. Nucl. Instr. Phys. Res. A 580, 1414–1422 (2007).

25. Marcus, Y. Effect of ions on the structure of water. Pure Appl. Chem. 82, 1889–1899 (2010).

26. Longeville, S. & Lechner, R. E. Light and heavy water dynamics.Physica B 276, 534–535 (2000).

27. Hong, L., Smolin, N. & Smith, J. C. de Gennes narrowing describes the relative motion of protein domains. Phys. Rev. Lett. 112, 158102 (2014).

28. Wuttke, J. et al. Neutron and light scattering study of supercooled glycerol. Phys. Rev. Lett. 72, 3052–3055 (1994).

29. Schober, H., Tölle, A., Renker, B., Heid, R. & Gompf, F. Microscopic dynamics of AC60 compounds in the plastic, polymer, and dimer phases investigated by inelastic neutron scattering. Phys. Rev. B 56, 5937–5950 (1997).

30. Seydel, T., Wiegart, L., Juranyi, F., Struth, B. & Schober, H. Unaffected microscopic dynamics of macroscopically arrested water in dilute clay gels. Phys. Rev. E 78, 061403 (2008).

31. Hill, A. E., Shachar-Hill, B. & Shachar-Hill, Y. What are Aquaporins for? J. Memb. Biol. 197, 1–31 (2004).

32. Tanghe, A., Van Dijck, P. & Thevelein, J. M. Why do microorganisms have aquaporins? Trends Microbiol. 14, 78–85 (2006).

33. Savage, D. F., Egea, P. F., Roles-Colmenares, Y., O'Connell, J. D. &Stroud, R. M. Architecture and selectivity in Aquaporins: 2.5 Å X-Ray structure of Aquaporin Z. PLOS Biology 1, 334–340 (2003).

34. Tajkhorshid, E. et al. Control of the selectivity of the Aquaporin water channel family by global orientational tuning. Science 29, 525–530 (2002).

35. Silantes, GmbH. Rich growth media solution labelled with stable isotopes (2H,13C,15N). http://www.silantes.com/media_solution.pdf (Date of access: 26/10/2015).

36. Jacrot, B. The study of biological structures by neutron scattering from solution. Rep. Prog. Phys. 39, 911–953 (1976).

37. van Heijkamp, L. F. et al. Spin-echo small angle neutron scattering analysis of liposomes and bacteria. J. Phys. Conf. Ser. 247, 012016 (2010).

38. Efimova, Y. F., Haemers, S., Wierczinski, B., Norde, W. & van Well, A. A. Stability of globular proteins in H2O and D2O. Biopolymers 85, 264–273 (2006).

39. Richard, D. Ferrand, M. & Kearley, G. J. Analysis and Visualisation of Neutron-Scattering Data. J. Neutron Research 4, 33–39 (1996); LAMP, the Large Array Manipulation Program.http://www.ill.eu/data_treat/lamp/the-lamp-book/. (Date of access: 26/10/2015).

40. Chen, S. H., Teixeira. J. & Nicklow, R. Incoherent quasielastic neutron scattering from water in supercooled regime. Phys. Rev. A 26, 3477–3482 (1982).

3

EXPERIMENTAL METHODS IN THE STUDY OF NEUTRON SCATTERING AT SMALL ANGLES

Cristian A. Dragolici[1]

[1]Horia Hulubei National Institute for R&D in Physics and Nuclear Engineering (IFIN-HH), Bucharest-Magurele, Romania

ABSTRACT

Scattering methods are powerful tools used for the examination of condensed matter. They offer the possibility to analyze particles without disturbing their natural environment. Small angle neutron scattering (SANS) can also be used to analyze solid and liquid systems, phase transformations, germination, growth flaws and defects, as well as generally any inhomogeneity occurring in a range of 10–1000 Å. The most common experimental SANS methods and patterns are reviewed and explained.

Keywords: SANS, methods, patterns

1. INTRODUCTION

1.1. Material Science Issues Accessible Through Small Angle Scattering Of X-Rays (Saxs) And Neutrons (Sans)

The fundamental properties of the neutron make it a powerful tool for materials science investigations. It is useful to introduce neutrons by comparison with X-rays because both probes are used for diffraction investigations, and more workers are familiar with X-ray methods. X-rays can be obtained with a generator tube or, in the past few years, at a synchrotron X-ray laboratory.

Neutrons for neutron beam research arise either from nuclear fission in reactors or from the spallation process. In the latter, accelerator-produced high-energy charged particles yield a large number of energetic neutrons from collisions with heavy-atom nuclei in the accelerator target.

X-ray diffraction was utilized for the first time in establishing the atomic structure of the crystals. Later, this method was used in other applications, extended to the study of imperfections in crystals, of the size of the crystallites, and even to the study of amorphous objects. From the classical discovery by Laue, in 1912, that X-rays are diffracted on their passage through crystals, the development of the applied domain of this technique was due to the subsequent progress of the X-ray diffraction theory and the improvement of experimental methods. The first observations on small angle X-ray scattering (SAXS) were performed in the beginning of the 1930s [1], and its domain was developed later.

In the following years, the small angle X-ray scattering technique has known to be an important development, proved by a great number of publications over the years. SAXS technique was developed from the necessity to observe long lattice spacing in crystals, in comparison with wavelengths of the X-rays used in structural analyses. Such distances are encountered in certain minerals and certain complex molecules such as polymers or proteins. For example, in the study of macromolecular crystals, the systems for X-ray diffraction should be extended to include the very small angles. The fundamental equation which describes the X-ray diffraction in the crystalline substance is $\boldsymbol{\lambda} = 2d\sin \boldsymbol{\theta}$, which shows that the diffraction angle θ inversely varies with the separation distance of the crystalline lattice planes (see Fig. 1) which generates the considered diffraction maximum.

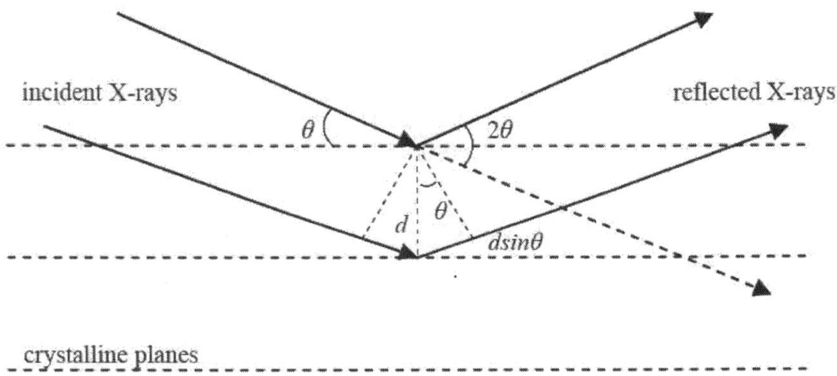

Figure 1. Bragg diffraction of the X-rays.

In ordinary crystals, especially those of inorganic matter, most inter-lattice distances observed are on the same magnitude with the wavelength of utilized X-rays; therefore the used θ angles are quite large. For example, let us consider the radiation Cu k_α and an inter-lattice distance of 100 Å, which would result in a diffraction θ angle of 0.45°; for a distance of 1000 Å, θ would be 0.045° or 2.6'. All these examples had shown the importance of SAXS techniques in biology, biochemistry, etc. Using X-rays with long wavelength to obtain larger diffraction angles for a specific inter-lattice distance is strictly limited by the variation of the material X-ray linear absorption coefficient with radiation wavelength. Because X-rays are strongly absorbed by the traversed substance, it also became necessary for the utilization of other particles with higher penetrability due to their weak interaction with the matter they are passing through. Elsasser first observed in 1936 that neutron movement could be determined by wave mechanism, and later two experiments were conducted by Halban and Preiswerk and Mitchell and Powers, respectively, which validated this theory [2]. These experiments were performed using source emitting neutrons of different energies, without considering collimation or monochromation of that beam. However, all these conditions were sufficient to emphasize the neutron diffraction but without providing quantitative data. The development of nuclear reactors made possible the availability of neutrons in a very large number to allow collimation in beams, and separation by energy in a narrow band. This was the beginning of the neutron diffraction technique, very similar to X-ray diffraction.

The first device of this type, initially called "neutron spectrometer," was built for the first time in 1945 at Argonne National Laboratory in USA. Since then, many instruments with different shapes and conception were built around the world, and they are mostly known as "neutron diffractometer." The investigation of condensed matter with the help of neutrons is not a substitute for classical X-ray diffractions, but it is a completion of these techniques, because the special nature of neutron interaction with the substance allows the neutron diffraction to provide information that cannot be revealed by other means. The spectrum of neutrons emitted by a reactor is dependent on the moderator temperature, and in most reactors, the spectrum peak is around 1.5 Å. This spectrum is Maxwellian and has a maximum which depends on the temperature. Therefore, for T = 273 K, λ_{max} is around 1.55 Å, and for T = 373 K λ_{max} is 1.33 Å. Using "hot sources," it is possible to move the maximum of the neutron spectrum to short wavelengths up to 0.4 Å. In a similar manner, by cooling the moderator, the maximum of the neutron spectrum can be moved to long wavelengths. We can observe that neutron beams, as spectrum broadening, become comparable with X-ray beams, where the normal range of wavelengths is from 2.2 Å (Cr K_α) to 0.56 Å (Ag K_α), thereby having the advantage of being less absorbed by the matter than X-rays. In the last years, in the competition between X-rays and neutrons, the synchrotron sources have been developed without compromising on the advantages of neutron diffraction. Today, the progress of X-rays and thermal neutron radiation sources makes it possible to use diffraction experiment radiations with λ up to 10 Å, and the development of the technology to obtain crystals with "pre-established inter-lattice" distances assures the conditions to extract monochromatic beams from the radiation emitted by the source. In their competition with X-rays, the neutrons have the huge advantage of being less absorbed by the matter in comparison with X-rays, which allow them to be used in the investigation of not only volume effects but also surface effects. In the case of neutrons, there are two ways of interaction with condensed matter, with the nuclei or with the atomic magnetic moments of the investigated material. Over the years, several methods were developed to investigate the partial order of analyzed systems. Among these methods, the technique of small angle neutron scattering is the most valuable. Technically speaking, the term "small angle scattering" (SAS) does not refer always to small diffraction angles but to small values of the quantity$Q$ which is the scattering vector, $\vec{Q}=\vec{k}-\vec{k}_0$, where $\vec{k}_0$ and $\breve{\vec{k}}$ are the wave vectors of the incident and scattered radiations. For elastic scattering $Q = (4\pi/\lambda)\sin\theta$, where θ is half of the scattering angle and λ is the radiation wavelength. In the angular range of SAS, the interested intensity is produced by inhomogeneities extended from 10 to 1000 Å. The quantity D of these inhomogeneities determines the most interesting angular range [3]:

$$\frac{1}{D} < Q < \frac{10}{D}$$

There are many processes that lead to scattering length variations, extended to zones up to 100 Å. Among these, there are the processes in solid phase namely aging, precipitating processes in solid phase, repeated thermal shock, neutron irradiation, the processes in mixed systems, for example the catalysts, where the two coexisting solid crystalline components are of enormous interest, as well as the ferofluids or the polymers in which neither one component is crystalline. The application of SAXS and SANS to material science has its origin from the experiments performed by Guinier in 1938, to study different types of aluminum alloys with various components such as Al–Cu and Al–Ag. He proved the existence of extremely small precipitate zones (so-called Guinier–Preston (GP) zones) as the first step in solidification of the precipitate of these alloys [3]. The GP zones represent submicroscopic regions of the matrix in which are produced a concentration of the initial solvate element. From one point of view, the GP zones could be considered as pre-precipitates because their formation represents a preliminary stage of precipitations set up from ultra-saturated matrix, but from another point of view it could be considered "clusters," in the sense that represent preferential associations of atoms.

In Al–Cu alloys, it has a diameter of about 80 Å and a thickness of 3–6 Å. With an average composition of 90% Cu, it comes out that the GP zones in these alloys are predominant constituted from Cu atoms. The physical nature and structural characteristics of the formed precipitates are dependent on temperature and heating duration because the precipitates are formed by germination and growing up processes, based on diffusion phenomena. The most important modification of alloy proprieties occur in the initial stages of the aging, with low temperatures and short heating durations when microstructural transformations are to a scale size under the resolution limit of optical microscopy (submicron precipitates). All these phenomena could be emphasized by SAXS and SANS methods. Regrouping of various atoms species during the precipitation process modifies the dispersion strength of X-rays or neutrons in different zones of aged alloy, due to the existence of an important density difference between the precipitated zones and the rest of alloy matrix. Among the structural processes produced through thermal treatment, which is applied to solid crystalline materials, the spinodal transition is something special [3]. The first phases of coherent precipitation took place through phase separation inside the crystalline lattice. Phase separation could start only through a nucleation process followed by nuclei growing. The germination of a new phase in a solid solution needs that in some regions the solid solution should considerably modify its composition. To be transformed into stable germs, these regions should have a specific dimension (about 10 Å). In these conditions, the interface between the germ and the solid solution (matrix) shows a distinct structural discontinuity, and therefore it has a free positive energy which inhibits the germination. This germination difficulty is more evident in spinodal decomposition. In this case, the matrix regions, where the composition is modified, became more extended than classical germs and are named clusters. The interface between the matrix and the regions with modified composition became diffusive and does not present anymore a distinct structural discontinuity characterized by a positive free energy. To produce regions with modified composition, extended on large volumes, long-range composition fluctuations are necessary that would tend to decompose the original solid solution. The probability of these long-range composition fluctuations is much more increased when the temperature of original solid solution is situated close to the alloy "spinodal temperature." Above the spinodal temperature T_s(where the second derivative of the Gibbs free energy for the solid solution is positive), the supersaturated solid solution is metastable. Below the T_s spinodal temperature, the second derivative of the Gibbs free energy is negative, resulting in an unstable solid solution, which separates immediately by spinodal decomposition. Concentration fluctuations of spinodal transition are detected by their geometrical parameter values, in SAS methods structural observation domain. By these methods, it can be defined as a wavelength λ_c and obviously its reciprocal $Q = 2\pi/\lambda_c$ that depend on different factors as temperature difference $(T_s - T)$ or energy gradient. Cahn provided a quantitative description of the primary phase's kinetics of spinodal decomposition, in his linear approximation theory. In accordance with this theory, the growing rate has a maximum for the wavelength $\lambda_m = 2^{1/2}\lambda_c$. Thus both wave vectors Q_c and Q_m can be experimentally deduced by SAS because the intensity $I(Q)$ is proportional with the amplitude square of the suitable concentration waves. SAS methods provide quantitative information not only about particle dimensions but also about volume fraction of the second phase as well as metastable mixture limits. Therefore, the idea of using SANS as a nondestructive testing method for the industry came to reality. Because neutron absorption in the substance is very low, machine components of relatively large thickness can be easily investigated by SANS. The initial steps in this direction began in Italy by a research group from Fiat Corporation. They built a cold neutron source for different material studies. They began in 1974 when Pizzi and coworkers performed several tests on various materials used in machine components and spare parts [3]. For example, they tested the palettes of a turbine that works for a very long time in a power plant. The base material was a Ni alloy, with the particle dimensions between 200 and 300 Å, which confer some mechanical properties to that alloy. After 16,000 of hours working time, they noticed a radial growth of the constituent particles, in the middle and top of the palettes, reaching 900 Å when the breaking occurred after approximately 60,000 working hours. Similar effects were observed in the aviation turbine palettes, where the heavy working conditions affected the particles situated mostly in the middle of the palettes. Other tests were carried out on different machine parts built of heterogeneous materials subject to cyclic deformations or having areas affected by strong heating (welding zones). The conclusion was that any change in the dimension of constituent particles (up to 1000 Å) could be easily observed using the SANS technique. Another field of SAS applicability is the study of flows (voids) appearing in some materials irradiated with neutrons [3]. The components of nuclear reactors are subject to massive exposure to high rapid neutron fluxes, therefore resulting in the necessity of the determination and study of defects that might appear in those materials, having effects on the modification of initial proprieties. The stress yielding of the material is correlated with the occurrence of a high concentration of micropores having a diameter between 200 and 1000 Å. Pores having smaller dimensions ($\leq$ 25 Å) are usually present in a very large number and influence the material density. The intermediary sized pores give the gas permeability and the stability to irradiation of the covering layers. All these become visible by SAS technique utilization. SANS is a very useful instrument in the study of microscopic distribution of the magnetic field and the static interaction of spins from different magnetic system as well. Magnetic scattering is caused by the interaction between the neutron magnetic moment and the local magnetization of the sample. If this occurs at a small Q, it will show the magnetization fluctuations adequate for large distances in the studied sample and also the magnetic periodicities extended on large distances.

Another important field of study is offered by complex disordered materials such as ceramics, clays, cement, or glasses. Among these, the importance of cement in modern society cannot be underestimated. Cement is found in concrete structures everywhere such as buildings, roads, bridges, dams, and even for conditioning radioactive wastes. There is no escape from the impact of cement in our everyday life. Due to its great importance in the following, we will point only on its applications. Cement, as it is commonly known, is a mixture of compounds made by burning limestone and clay together at very high temperatures ranging from 1400 to 1600°C. Water is the key ingredient, which when mixed with cement forms a paste that evolves as a hard product. The water causes the hardening of the cement by a process called hydration. Hydration is a chemical reaction in which the major compounds in cement form chemical bonds with water molecules to yield hydrates or hydration products.

The heterogeneous nature and chemical complexity of the cement make the characterization of this system difficult. In spite of the inherent difficulties of the system, considerable understanding of the nature of cement and concrete has been obtained through the use of a wide variety of tools.

Studies of the cement pore liquid as a function of time, X-ray diffraction investigations, electron microscope, and NMR experiments have all contributed to a better understanding of the behavior of this material.

The greater penetrating power of neutrons (in comparison to X-rays), the opportunities for in situ measurement of hydrating pastes, the ability to use very long wavelength incident neutrons and the unique opportunities to change contrast offered by D_2O–H_2O exchange make SANS attractive as a means to probe the internal microstructure of cement pastes. The issue to be resolved, however, is the interpretation of the small angle scattering data. While the actual measurements are generally straightforward, the complexity of phases, contrasts, object morphologies, and size ranges forms a barrier to simple interpretations.

Small angle neutron scattering was first used to study the microstructure of cements in the early 1980s by Allen and coworkers [4]. They measured the small angle neutron scattering from a set of cement specimens hydrated at different water to cement ratios. They also examined the effect of soaking the specimen in heavy water and the change in SANS obtained when the specimen was dried. In their analysis of the SANS data, they assumed a simplified model of the cement hydration reactions. In this model, the hydrated compounds C_6AFH_{12}, C_4AH_{13}, C_3ACSH_{12}, and $C_3S_2H_{2.5}$ and the hydroxides $Mg(OH)_2$, $Ca(OH)_2$, NaOH, and KOH in appropriate amounts were the products of cement hydration. Each of these compounds represented a different volume fraction of the cement paste. The strength of the SANS signal from the specimen in the small Q Guinier region was directly proportional to the square of the scattering length density difference (the contrast, see Equation 18) and to the total volume of the scattering particles. If the scattering in the two-phase model is to be interpreted as between the matrix $C_3S_2H_{2.5}$ and pores in the $C_3S_2H_{2.5}$ gel, it matters whether the pores are filled with light water, heavy water, or with air. The contrast between $C_3S_2H_{2.5}$ and water will vanish where the scattering length density curves intersect, at approximately 60% D_2O. If SANS data from cement paste are to be interpreted on the basis of a two-phase model, one issue is the identification of the appropriate phases. While the contrast is proportional to the SANS intensity, the detailed shape of the SANS curves will depend on the size distribution and shape of the scattering objects.

The SANS is presumed to be caused by a size distribution of objects of a known shape. To calculate the form factor for objects of known shape we can use Equation 17.

The parameters describing the object and its size distribution are then varied to minimize the difference between the calculated scattering and the data. Spheres, cylinders, and disks were considered by Allen and coworkers, and it was ultimately concluded that the SANS from their cement paste specimens was created by a distribution of water-filled spherical pores approximately 5 nm in diameter and a smaller component of pores with diameters of about 10 nm. SANS from the specimen that had been dried and not rewetted was interpreted as the one due to scattering from a much broader distribution of pore sizes with a peak of about 5 nm but extending to much larger diameters.

SANS from the specimens immersed in D_2O showed that the heavy water rapidly exchanged with the water in the pores and with the water in the surrounding C–S–H gel. The total porosity was of the order 1% of the volume.

All these are only a few examples that illustrate the large field of SAS application, where the immediate observation of the phenomena produced in materials lead to solving many practical problems and to extend the theoretical knowledge as well.

2. PHYSICAL PRINCIPLE OF SANS

SANS is a method used to obtain information regarding the shape, dimensions, and internal configuration of the zones with scattering density significantly different from the average value, having relatively small dimensions (up to 1000 Å), unevenly distributed in some environment, where the distance arrangement is present or not. In a considered sample, all scattering centers participate in the scattering process. The scattering centers could be atoms, molecules, or particles of the analyzed system. System response could be a spatial distribution of scattered radiation reducible to a succession of maximums, determined by the existence of distance arrangement in a material, or a succession of overlapping maximums over a continuous distribution when the distance arrangement is accompanied by disordered processes, such as amorphization, precipitation in solid phase, flaws, etc., or by a continuous distribution which present one or two isolated maximums like in the case of liquids (in this case the maximum is owed to the next close vicinity of each scattering center) or a continuous curve in case of SANS. Because X-ray scattering is similar to neutron scattering, in the following presentation I will start with the theoretical principles of SAXS accordingly adapted for SANS. It is well known that a diffraction image of a sample could be easily described in terms of reciprocal space or Fourier space [1]. If we consider $\rho(\vec{x})\rho(x\rightarrow)$ as the electronic density of the scattering body in a point defined by the vector $\vec{x}\, x\rightarrow$, then $A(\vec{Q})A(Q\rightarrow)$ is defined as the transform of $\rho(\vec{x})\rho(x\rightarrow)$ in the point defined by the vector $\vec{Q}\, Q\rightarrow$ in the reciprocal space, and it is given by

$$A(\vec{Q}) = \int \rho(\vec{x}) \exp(-i\vec{Q}\cdot\vec{x})\, d\vec{x} \tag{1}$$

The X-ray diffraction theory is based on the fact that $A(\vec{Q})A(Q\rightarrow)$ represents the amplitude of scattered radiation when $\vec{Q}\, Q\rightarrow$ is defined as

$$\vec{Q} = \left(\frac{2\pi}{\lambda}\right)(\vec{s} - \vec{s}_0) \tag{2}$$

where λ is the wavelength of the radiation, $\vec{s}\, 0s\rightarrow 0$ and $\vec{s}\, s\rightarrow$ are the unit vectors on the incident radiation direction and scattered radiation direction, respectively. The magnitude of $\vec{Q}\, Q\rightarrow$ is equal to $(4\pi/\lambda)\sin\theta$, where $2\theta = \theta_s$ is the scattering angle (the angle between incident ray and scattered ray). Therefore, very small scattering angles correspond to small values of the Q quantity. Analyzing Equation (1), we find that the intensity of the scattered radiation observed for a specific value of Q⃗ Q→ is equal to the square value of A(Q⃗)A(Q→), where A(Q⃗)A(Q→) is the suitable component of Q⃗ Q→ in the Fourier serial development of ρ(x⃗)ρ(x→). For small values of Q⃗ Q→ that are produced at very small angles, the terms in ρ(x⃗)ρ(x→) that define the quantity A(Q⃗)A(Q→) give a periodicity of $x = 2\pi/Q$, which is large enough in comparison with X-ray wavelength. These general considerations show that the very small diffraction angle (smaller than a few degrees) provides information on the structure of the matter to a relatively large scale in comparison with X-ray wavelength. It was experimentally observed that some samples generate a continuous and intense scattering at angles under 2°, without producing the normal diffraction effects from the usual X-ray experiments. This was first time noticed in specific types of fine carbon, black carbon powders, and different types of other substances, all having in common the fine particles of submicroscopic sizes. Later, it was concluded that continuous scattering around the direct beam is owed to the existence of the matter composed of small particles or, more general, to the existence of heterogeneities in the analyzed substance that can have dimensions from a few tens to a few hundred times the wavelength of the X-ray. The qualitative description of the central scattering, due to the presence of very small particles, could be done by analogy with the well-known light diffraction phenomena, which lead to the appearance of a halo when the light ray traverses a powder with grains having the dimensions 10 times larger than the wavelength of the light radiation.

Let us consider a particle bathed in an X-ray beam; thus all the electrons are wave scattering sources. If the scattering direction is the same as the incident ray, we can say that the scattered rays are all in phase and, if the scattering angle increases, the phase difference between the different scattered waves will also increase. Therefore, the amplitude of the resulting scattered wave will have values bigger or smaller in accordance only with the existing phase differences. This happens for a scattering angle about $2\theta = \lambda/D$, D being the average size of the particle, proving in this way that the study of central scattering offers a method to obtain the particle dimensions. This method is applicable only to the particles with dimensions between some specific limits. If D is too big, the scattering is limited to angles so small that are experimentally inaccessible. If *D* is too small, of the order of a few wavelengths, the scattering is too broad and too weak to be observed. To show more precisely which factors depend on the small angle scattering, let us consider a small particle of electronic density ρ(x⃗)ρ(x→) . Let us define a form factor s(x⃗)s(x→) of this particle, which has a value of 1 when x⃗ x→ vector is inside the particle, and a value of 0 when x⃗ x→ is outside the particle. Thus, according to Equation (1), the amplitude of the scattered radiation from this particle will be

$$A_1\left(\vec{Q}\right)=\int \rho\left(\vec{x}\right)s\left(\vec{x}\right)\exp\left(-i\vec{Q}\cdot\vec{x}\right)d\vec{x} \tag{3}$$

If A(Q⃗)A(Q→) and S(Q⃗)S(Q→) are the Fourier transforms of ρ(x⃗) ρ(x→) and s(x⃗)s(x→), respectively, then

$$A_1\left(\vec{Q}\right)=A\left(\vec{Q}\right)\cdot S\left(\vec{Q}\right) \tag{4}$$

This can be demonstrated in the following manner. Let us consider the functions

$$a\left(\vec{x}\right)=\rho\left(\vec{x}\right)^{*}s\left(\vec{x}\right)\text{ and }a\left(\vec{x}\right)=\int \rho\left(\vec{y}\right)\cdot s\left(\vec{x}-\vec{y}\right)dv_y$$

Thus, the Fourier transform of a(x⃗)a(x→) is

A1(Q⃗)A1(Q→) ;hence

$$A_1\left(\vec{Q}\right)=\int a\left(\vec{x}\right)\exp\left(2\pi i\vec{Q}\cdot\vec{x}\right)dv_x=\iint \rho\left(\vec{y}\right)s\left(\vec{x}-\vec{y}\right)\exp\left(2\pi i\vec{Q}\cdot\vec{x}\right)dv_x dv_y$$

Using the notations *Y* = *y*, *Z* = *x* – *y* we replace the elementary differentials *dvxdvy* with *dvYdvZ* and to obtain

$$A_1\left(\vec{Q}\right)=\iint \rho\left(\vec{y}\right)s\left(\vec{Z}\right)\exp\left[2\pi i\vec{Q}\cdot(\vec{Y}+\vec{Z})\right]dv_Y dv_Z=$$
$$=\int \rho\left(\vec{Y}\right)\exp\left(2\pi i\vec{Q}\cdot\vec{Y}\right)dv_Y\cdot\int s\left(\vec{Z}\right)\exp\left(2\pi i\vec{Q}\cdot\vec{Z}\right)dv_Z=A\left(\vec{Q}\right)\cdot S\left(\vec{Q}\right)$$
$$A_1\left(\vec{Q}\right)=A\left(\vec{Q}\right)\cdot S\left(\vec{Q}\right)$$

Giving the dimensions of the region where s(x⃗)s(x→) is different from zero, its transform S(Q⃗)S(Q→) is completely determinated and, if the particle has dimensions from some few tens to a few hundreds of atomic diameters, $S(\vec{Q})S(Q\rightarrow)$ will be different from zero only for very small values of $\vec{Q}\ Q\rightarrow$. Now, let us consider the function $A(\vec{Q})A(Q\rightarrow)$. If we suppose that the sample has a constant electronic density $\rho(\vec{x})=k\rho(x\rightarrow)=k$, then the transform $A(\vec{Q})A(Q\rightarrow)$ will behave like a Dirac delta function, being zero overall with the exception of the point $\vec{Q}\ Q\rightarrow = 0$, where it is infinite and $\int \delta(\vec{x})d\vec{x}=1.$. We can conclude that small angle scattering is the method to study the dimensions of the system constituent particles, but the supposition of the existence of the identical particles separated by long distances between them is not quite accurate in the case of real samples. Therefore, the generalization of the theory is needed to take into consideration the diversity of particle sizes and shapes as well as the effect of particle close packing. Without considering the particles, a mathematical expression should be obtained for the scattered intensity in the vicinity of the center that should take into account the electronic density in all the points of the sample. Further, we will discuss only the process of single scattering, neglecting the multiple scattering phenomena and Compton scattering. We will consider an X-ray beam, sufficiently large to irradiate a large number of particles, but narrow enough in comparison with the sample-detector distance, to be like a ray in the experimental device geometry. The amplitude of scattered radiation from the *Pn* point (Fig. 2) will depend on the scattering factor *un* and the direction defined by the unit vector

$\vec{s}\ s\rightarrow$

$$A_n = A_e u_n \exp\left[-i\frac{2\pi}{\lambda}\left(\vec{s}-\vec{s}_0\right)\cdot\left(\overrightarrow{OP_n}\right)\right] \quad (5)$$

where *Ae* is the amplitude scattered by a single electron in the same conditions, *O* is an arbitrary origin that serves to describe the pathway difference between different rays, and $\vec{s}\ 0s\rightarrow 0$ is the unit vector defining the incident radiation direction.

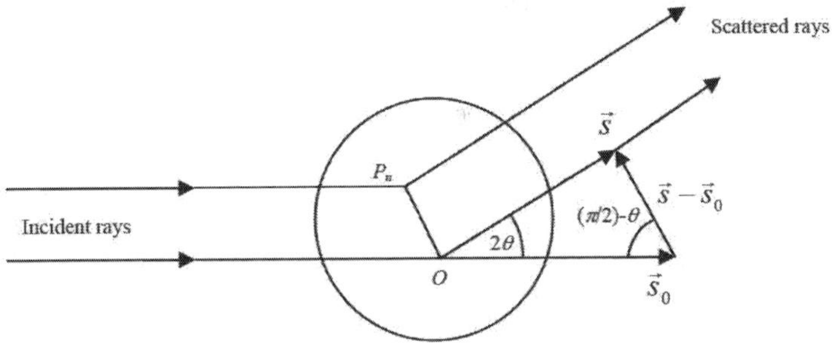

Figure 2. Diffraction by a scattering body.

The total amplitude of scattered radiation will then be

$$A(\vec{Q}) = \sum_n A_n = A_e(\vec{Q}) \sum_n u_n \exp\left(-i\vec{Q}\cdot\overrightarrow{OP_n}\right) \tag{6}$$

and the scattered intensity, the product between the amplitude *A,* and its complex conjugate A^*, will be

$$I(\vec{Q}) = \left|A_e(\vec{Q})\right|^2 \sum_n \sum_j u_n u_j \exp\left[-i\vec{Q}\cdot\left(\overrightarrow{OP_n} - \overrightarrow{OP_j}\right)\right] \tag{7}$$

The intensity scattered by an electron is

$$I_e(Q) = A_e^2(Q) = A_0^2 \frac{e^4}{m^2c^4}\cdot\frac{1}{r^2}\sin 2\theta = I_0 \frac{e^4}{m^2c^4}\cdot\frac{1}{r^2}\sin 2\theta \tag{8}$$

which is a three-variable function, the incident intensity, the scattering angle, and the distance of observation; I_0 is the intensity of incident beam; and *r* is the distance from the particle to detector. If the analyzed environment has a center of symmetry, and this center is placed in the origin *O*, then to each OPn−→-OPn→ vector will correspond another vector −OPn−→−-OPn→. Thus, we will obtain a simplified expression for the scattered amplitude:

$$A(\vec{Q}) = \sum_n A_n = A_e(\vec{Q}) \sum_n u_n \cos\left(\vec{Q}\cdot\overrightarrow{OP_n}\right) \tag{9}$$

We define the structure factor of the environment as the report between the total scattered amplitude and the radiation amplitude scattered by a single electron in the same conditions:

$$F(\vec{Q}) = \frac{\sum_n A_n(\vec{Q})}{A_e(\vec{Q})} = \sum_n u_n \cos\left(\vec{Q} \cdot \overrightarrow{OP_n}\right) \tag{10}$$

The scattered intensity will then be

$$I(\vec{Q}) = I_e(\vec{Q}) \left[\sum_n u_n \cos(\vec{Q} \cdot \overrightarrow{OP_n}) \right]^2 = I_e(\vec{Q}) F^2(\vec{Q}) \tag{11}$$

The term "*Pn* point" was used to define the structure of a particle. If we consider a large particle, the base element in its description is the atom; in this case, the point *Pn* refers to the center of the *n*th atom, and the scattering factor *un* is the scattering factor of this *n*th atom. In the interested angular domain for SAS, un(Q→)un(Q→) can be considered as a constant equal to *un*(0). If we consider a small region, then *Pn* point refers to a small volume element that surrounds it. The scattering factor *un* is then equal to *ρndvn,* where *ρn* is the electronic density of the particle in the vicinity of *Pn* point, and *dvn* is the volume of considered element. Generally, it is convenient to describe the particle structure taking into account enough small elements to consider the scattering factors of these elements to be constant, independent of scattering angle over the whole domain where the structure factor of the considered particle is different from zero. This theory of small angle X-ray scattering can be adapted very easily to neutrons.

Small angle neutron scattering is used where the X-rays cannot provide the desired information either of the lack of scattering contrast or of severe absorption of the studied material. The choice of thermal neutrons as a means of substance investigation was owed to their wavelengths and their energies corresponding to the inter atomic distances and excitation energies of condensed matter. Neutron absorption from the matter is very low, thus resulting in the samples that can have thicknesses larger than in the X-rays case which are strongly absorbed with an increase in the volume of analyzed samples. We can say that SANS allows the investigation of materials in various conditions, such as in containers, furnaces, cryostats, etc. Due to its magnetic moment, the neutron provides unique possibilities to the study of magnetic structures, magnetic moment distribution, and magnetic excitations. Being a nuclear propriety, the nuclear scattering amplitude can be considerably different between various isotopes of a specific chemical species. For example, the big difference between the coherent scattering lengths of hydrogen and deuterium lead to the usage of phase contrast in the study of hydrogenate materials, allowing a good resolution in analyzing polymers and biological substances in general. Because thermal neutrons interact very low with the matter, this interaction can be theoretically treated on the base of first Born approximation [5, 6]. We will show further that the variation of the scattering density is very important in the study of condensed matter through SANS. In the following, we will consider the "static approximation"; the scattering process is produced only at nuclear level on fixed targets letting those targets unchanged after the collision, and without considering the polarization effects.

Thus, we can write the coherent elastic scattering differential cross section on a single atom:

$$\frac{d\sigma}{d\Omega} = \frac{1}{N}\left|\sum_{R} b_R \exp\left(i\vec{Q}\cdot\vec{R}\right)\right|^2 \tag{12}$$

where *N* is the number of scattering nuclei exposed to the beam, and *bR* is the coherent scattering length of the chemical species that occupy a specific place determined by the position vector R⃗ R→. If we replace *bR* through an average local scattering length density ρb(r⃗)ρb(r→), where r⃗ r→ is a continuously variable position vector, we can write

$$\frac{d\sigma}{d\Omega} = \frac{1}{N}\left|\int_{V} \rho_b\left(\vec{r}\right)\exp\left(i\vec{Q}\cdot\vec{r}\right)d^3\vec{r}\right|^2 \tag{13}$$

where the integration is performed over the whole sample volume *V*.

The scattering length density $\rho b(\vec{r})\rho b(r\rightarrow)$ can now vary on distances of the order of d_{min}, where $d_{min} \cong \pi/Q_{max}$, Q_{max} being the maximal value accessible by experiment. Thus, we can write that

$$\rho_b(\vec{r}) = \Delta\rho_b(\vec{r}) + \bar{\rho}_b \tag{14}$$

where ρb is averaged on volumes larger than the instrument resolution volume, determined by the minimum observed value of Q. Therefore, the cross section that contains the useful information about the studied system is dependent on

$\Delta\rho b(\vec{r})\Delta\rho b(r\rightarrow)$. By substituting Equation (14) in Equation (13) we obtain

$$\frac{d\sigma}{d\Omega} = \frac{1}{N}\left|\int_V \left[\rho_b(\vec{r}) - \overline{\rho_b}\right]\exp\left(i\vec{Q}\cdot\vec{r}\right)d^3\vec{r}\right|^2 \tag{15}$$

For a specific distribution of scattering length, the cross section can be analytically calculated with the help of Equations (13) and (15). Because the cross section values are not known everywhere in reciprocal space, $\rho b(\vec{r})\rho b(r\rightarrow)$ determination by experimental measurements of $d\Sigma/d\Omega$ values is very hard to be carried out, or even impossible. Let us take a sample containing Np particles with scattering length densities $\rho bp = bp/vap$, where bp is the scattering length averaged on the particle volume, and vap is the atomic volume from the particle. Let us consider that these particles are packed in a homogeneous matrix having scattering length densities $\rho bm = bm/vam$.

Using Equation (13) we obtain

$$\frac{d\Sigma}{d\Omega}(\vec{Q}) = \frac{1}{N}\left(\rho_{bp} - \rho_{bm}\right)^2\left|\int_{V_t} \exp\left(i\vec{Q}\cdot\vec{r}\right)d^3\vec{r}\right|^2 \tag{16}$$

where the integral is extended over the volume Vt occupied by all the particles. In the most general case, this integral will contain spatial and orientation correlation elements between particles, and effects owed to the size distribution of the particles. Taking into account the form factor expression for a single particle

$$F_p\left(\vec{Q}\right)=\frac{1}{V_p}\int_{V_p}\exp\left(i\vec{Q}\cdot\vec{r}\right)d^3\vec{r} \tag{17}$$

where *Vp* is the particle volume so that $|F_p(0)|^2 = 1$. We will now consider *Np* identical particles for which we can write the following expression

$$\frac{d\Sigma}{d\Omega}\left(\vec{Q}\right)=\frac{V_p^2 N_p}{N}\left(\rho_{bp}-\rho_{bm}\right)^2\left|F_p\left(\vec{Q}\right)\right|^2 \tag{18}$$

that is valid in the case of binary systems, without contributions from the interferences between particles.

The interference term from Equation (16) that was neglected in Equation (18) is the Fourier transform Ψ(Q⃗)Ψ(Q→) of the correlation function:

$$\Psi\left(\vec{Q}\right)=\frac{1}{N_p}\sum_{i\neq j}\exp\left[i\vec{Q}\cdot(\vec{r}_{0i}-\vec{r}_{0j})\right] \tag{19}$$

where r⃗ 0ir→0i and r⃗ 0jr→0j are the position vectors of the particle centers denoted by *i* and *j*.The Ψ(Q⃗)Ψ(Q→) function will be zero for all the values of Q⃗ ≠0Q→≠0 if the distance distribution between the particles is completely random, approximately like in the case of very diluted systems. Equation (18) is valid for particles anisotropically oriented, but all are identically oriented. In most cases of random distribution of the orientations or in discrete but multiple orientations of anisotropic particles, suitable averages of $|F_p(\vec{Q})|^2$ are used. Guinier shows that at small values of *Q*, the scattering function

$S(\vec{Q}) = |F(\vec{Q})|^2$ for a specific physical system can be written as

$$S(\vec{Q}) = \exp\left(-Q^2 R_D^2\right) \tag{20}$$

where *RD* means the gyration radius which is a material parameter having the length dimension defined through

$$R_D^2 = \frac{1}{V_p} \int_{V_p} r_D^2 q(r_D) dr_D \tag{21}$$

where *q*(*rD*) is the geometrical cross section of the particle, through the length of a plane normal to a direction *D* and at a distance *rD* from the origin inside the particle. For a system of particles randomly oriented

$$S(\vec{Q}) = \exp\left(-Q^2 R_G^2 / 3\right) \tag{22}$$

with

$$R_G^2 = \frac{1}{V_p} \int_{V_p} r^2 d^3\vec{r} \tag{23}$$

$$S_S(QR_S) = \left[3\left(\sin(QR_S) - QR_S\cos(QR_S) / Q^3 R_S{}^3\right)\right]^2$$

to the term proportional with Q^4. The subsequent terms in the developments of the two series are in a good concordance, but the corresponding terms have the same sign. Guinier approximation is thus accepted over a large range of *QRG*(<1.2).

For rotational ellipsoids, the Guinier approximation coincides with the extension of scattering functions until Q^6. For other shapes of the particles, it might be necessary for an investigation of the small values of *Q* to be able to determine *RG* with sufficient accuracy. In the case of homogeneous particles, with pronounced margins and area *Ap*, Porod has proved that

$$\overline{S(\vec{Q})} \cong 2\pi A_p / V_p^2 Q^4 \tag{24}$$

This shows the average decrease in the scattering function at large Q. Q should be larger than the reciprocal value of the smallest dimension of the particle.

3. EXPERIMENTAL METHODS USED IN THE STUDY OF CONDENSED MATTER BY SANS

The goal of SANS method is to measure $I(Q)$ and to extract from that measurement the "structural" data regarding the studied system. The quantity Q is defined as $Q = \frac{4\pi \sin\theta_s}{\lambda} \approx \frac{2\pi}{\lambda}\theta$, if $\theta = \frac{\theta_s}{2}$ is small enough.

Therefore, to measure $I(Q)$, it is sufficient to measure $I(\theta)$ at constant λ or $I(\lambda)$ at constant θ. In the following, we will discuss both experimental alternatives.

3.1. The Method $I(\Theta)$ At Constant Λ

This method is based on monochromatic (λ = constant) beam extraction from the white thermal neutron spectrum provided by a research reactor. The thermal neutrons released from the reactor are first collimated with a Soller collimator (with plane parallel slits) and after that they are monochromatized with either a mechanical monochromator or a crystal. Monochromatization is the method that allows the extraction of desired neutrons from the neutron beam provided by the source; in other words, to obtain thermal neutrons with energies in a very narrow interval centered around a value E_0 of the energy. Neutron monochromation is based either on their corpuscular proprieties (mechanical monochromators) or on their wave proprieties (crystal monochromator). The mechanical monochromator is a cylinder endowed with helicoidal slits on the whole length of its generatrix.

When the cylinder is rotating with the angular speed ω, the neutrons with their motion direction parallel with the rotor axis would traverse the monochromator slits without modifying their distance to the slits walls only if they would have a specific speed v_0, directly correlated with ω and which is corresponding to the wavelength λ_0 of the neutrons transmitted by the monochromator. Another type of mechanical monochromator uses disks instead of a cylinder. Its functional principle is identical with the cylindrical monochromator, but having the difference that the place of helicoidal slits is taken by straight slits cut off on a rigid mounted disc on a horizontal axis, which rotates with an angular speedω. The disks are positioned on the axis having the slits displaced to provide a selective opening in accordance with the axis rotation speed ω, only to the neutrons moving with a specific speed v_0. Nevertheless, the most used monochromatization method is by using a crystal. Crystal utilization in spectrometer devices is based on the neutron diffraction phenomena.

Monochromatic neutron beam separation from the thermal neutron beam provided by the reactor is realized through its Bragg reflection on a monocrystal. If we note the distance between the crystalline planes with *dhkl* determined by the *h, k, l* indexes, and *θB* the neutron incident angle on these planes, then the reflected neutron wavelength is given by

$$n\lambda = 2d_{hkl}\sin\theta_B = \frac{nh}{\sqrt{2mE}}$$

the Bragg law

where *m* and *E* are the neutron mass and neutron energy, respectively, but *n* is the reflection order.

In other words, if a monocrystal is used to select neutrons with a specific wavelength λ that satisfies the Bragg relation, then the neutrons selected by the crystal would also have the wavelengths $\lambda/2$, $\lambda/3$,$\lambda/4$,..... etc., which means that reflections with a higher order take place, with n = 1, 2, 3,.... In the case of monochromators with crystal, the presence of high-order reflections could be a serious problem because sometimes, depending on the spectral distribution of the incident beam and the crystal structure factor, the reflection intensities become comparable and even greater than the first-order reflection intensity. In most neutron physics experiments, first-order reflection (n = 1) is used, but other reflections (high-order contaminations) should be eliminated by different means.

Analyzing the Bragg relation, we can see that by using of monochromator crystals, only monochromatic neutrons with the wavelength of λmax≤2dmaxλmax≤2dmax can be selected, where d_{max} is the maximum distance between the crystal planes. The neutrons having the wavelength $\lambda > 2d_{max}$ are not presenting Bragg reflections.

This observation suggests a simple way to eliminate the second-order reflections, utilized mostly when the monochromatic neutrons desired to be extracted from the reactor emergent beam have wavelengths relatively large, for example, in the range $\lambda > 6$ Å. It is sufficient to introduce a polycrystalline filter of convenient size in the pathway of neutron beam falling on the monochromator crystal, with its crystalline structure characterized by $d_{max} < \lambda_0/2$, where λ_0 is the wavelength of the neutrons to be extracted. There is another experimental solution to eliminate second-order reflections, for example, using a monochromator of special monocrystals with an internal composition that cancels the structural factor for some specific groups of crystalline indexes. Because the problem of high-order reflections is overpassing the goal of this chapter, we will not insist on this subject.

Frequently, the experimental pattern used to obtain monochromatic neutron beams in this way contains a Soller collimator which spatially delimitates the incident neutron beam coming out from the reactor, the monochromator crystal, and the second Soller collimator which delimitates the diffracted beam. Simultaneous rotations of the second collimator and the crystal with angular speeds staying in a ratio of 2:1 allows to extract monochromatic neutron beams of different wavelengths from the continuous spectrum of the incident thermal neutron beam. The monochromatic neutrons are recorded by a detector placed after the second collimator rotating in the same time with it.

The devices built on this principle, the so-called neutron spectrometer with crystal, allow the determination of total cross section by the measurement of sample transmission in accordance with neutron wavelength. To study the angular distribution of scattered neutrons by a certain target and implicitly to measure the differential cross sections, a pattern with a third Soller collimator between the sample and detector is used. To investigate the neutron scattering of different wavelengths, the whole system built from the second collimator, the studied sample, the third collimator, and the neutron detector should be rotated around the first fixed axis that is going through the middle of the monochromator crystal and together with it; the angular speed ratio of the crystal and the mentioned system should be 1/2. For a certain position of the monochromator crystal and the second collimator, the third collimator and the detector could be rotated around the second fixed axis (that is going through the middle of the sample) allowing the measurement of sample differential cross section for thermal neutrons having the wavelengths determined by the monochromator preceding the sample. This experimental pattern is named neutron spectrometer with two axes. Many experimental systems use optical patterns derived from the two axis spectrometer described earlier. Anticipating the discussion in the next section, we would mention that from the simple presentation of the optical pattern of considered spectrometer arises a specific limitation of its efficiency. Actually, to measure$I(Q)$ with sufficient precision in the optical pattern described before, it is necessary to increase the distances (R) between the studied sample and the detector, as much as possible; increasing this distance obviously will lead to decreasing the lower limit of the explored Q range. Because the intensity of the scattered beam varies with $1/R^2$, the obtained resolution gain is drastically affected by losing the luminosity of the experimental system. If we add to this inconvenience the construction of experimental devices having excessive dimensions, the continuous search for more efficient SANS optical patterns becomes explicable. One of the identified optical patterns which is considered most acceptable is by using two crystals, with the sample staying between them.

Two crystal systems were used as monochromators in spectrometric assemblies long time ago, first in X-ray physics and then in neutron physics. To discuss the utilization manner of two monocrystals systems in SANS, we have to reveal the optical characteristics of these systems. To simplify the explanation, we will limit our case to consider that the two monocrystals are ideal identical crystals, from the point of view of their internal crystalline structure and their cutting method (has the same active reflecting planes). A two-monocrystal system could work in two optical patterns: parallel (Fig. 3) and antiparallel (Fig. 4).

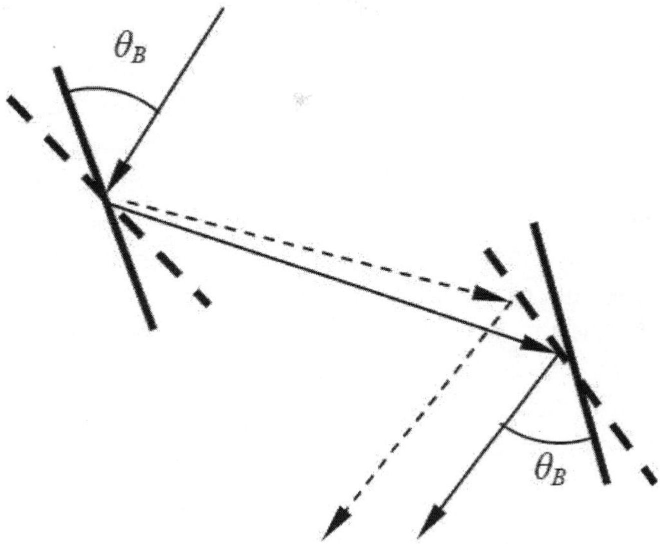

Figure 3. Parallel pattern.

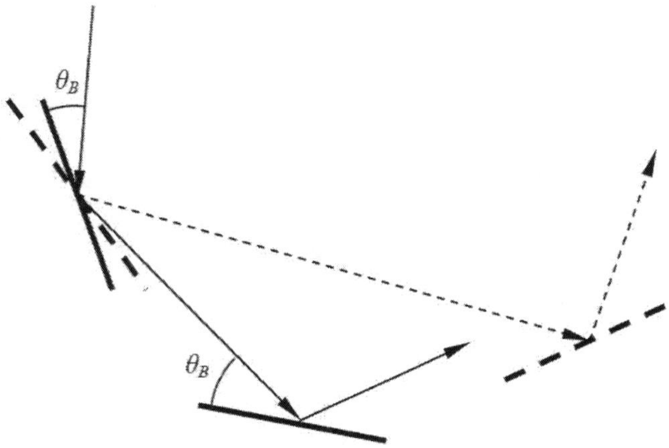

Figure 4. Antiparallel pattern.

In the parallel pattern, any selected wavelength from the incident polychromatic beam through Bragg reflection on the first crystal leads to the propagation of the double reflected beam in parallel direction with the incident beam. Therefore, this pattern is not dispersive (does not spatially separate the beams of different wavelengths). In the second optical pattern, the Bragg double reflected beam is propagated on a direction oriented toward the region containing the radiation source and is named antiparallel. To notice that the wavelength change of the selected radiation beam through double reflection is realized along two directions depending on the wavelengths, revealing the dispersive feature of the antiparallel pattern. In the case of the parallel pattern, the thermal neutron beam falling on the first monochromator crystal under the incident angle θB, it will be Bragg reflected on the second crystal, the double reflected ray making the same angle θB with the second crystal (Fig. 5).

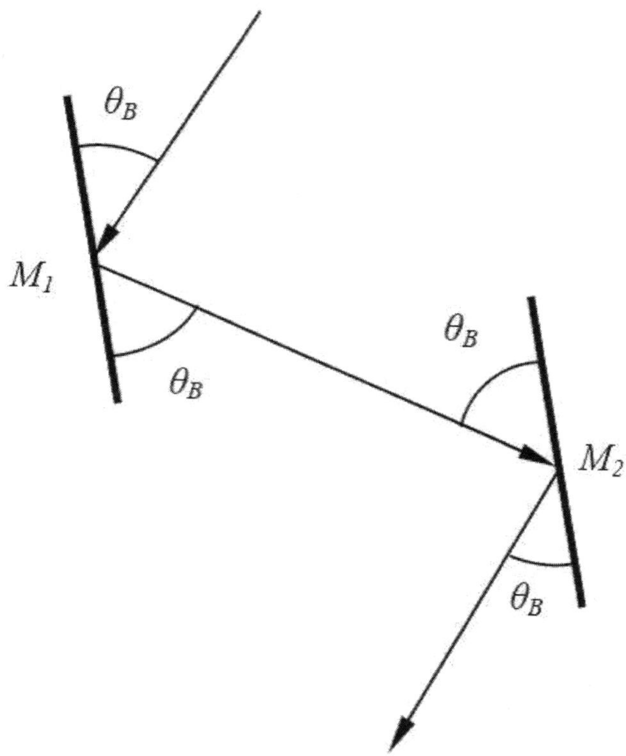

Figure 5. Double crystal parallel pattern.

If a sample is interposed between the two crystals, then the beam reflected by the first crystal will be scattered by the sample under the θS angle and will fall on the second crystal under the θ incident angle (Fig. 6).

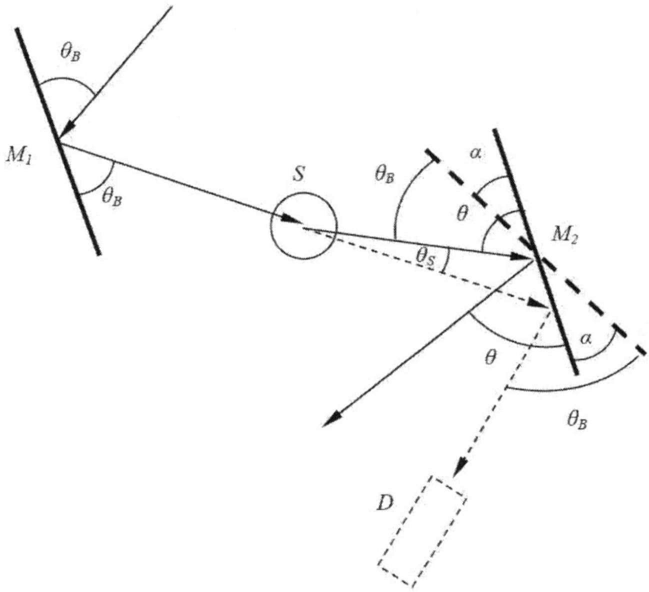

Figure 6. Three axes parallel pattern.

To detect the scattered beam, we have to rotate the second monochromator crystal together with the detector by an angle α; thus the Bragg relation is satisfied by obtaining the same incident θB angle for the sample scattered ray. Among these angles, we can write the following relations:

$$\theta - \alpha = \theta_B \Rightarrow \theta - \alpha = \theta_B$$
$$\theta_S + \theta_B = \theta \Rightarrow \theta - \theta_S = \theta_B \Rightarrow \alpha = \theta_S$$

Because the sample is scattering, the neutrons under different angles $\theta S(\theta S_1, \theta S_2, ..., \theta Sn)$ arise that the system composed from the second monochromator and the detector should be rotated in accordance with the suitable $\alpha(\alpha_1, \alpha_2,..., \alpha n)$ angles to be able to detect all scattered neutrons. Let us now consider the case of the antiparallel pattern. The thermal neutron beam falls on the first monochromator crystal under the incident θB angle and is reflected on the second monochromator crystal, the double reflected ray leaving the system under the same θB angle (Fig. 7).

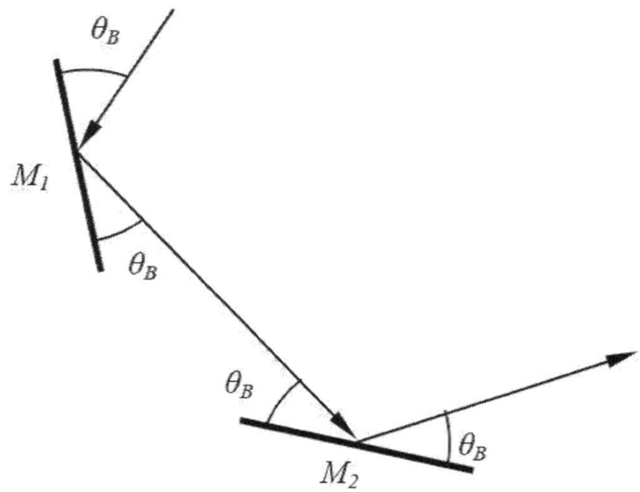

Figure 7. Double crystal antiparallel pattern.

If we have a sample between the two crystals, then the reflected beam by the first crystal will be scattered by the sample under a θS angle and will fall on the second crystal under the θ angle (Fig. 8).

The neutrons scattered by the sample S could be detected only by rotating the second crystal with the αangle, thereby reestablishing the Bragg law by remaking the initial incident θB angle. Also, we found that

$$\theta + \alpha = \theta_B \Rightarrow \theta + \alpha = \theta_B$$
$$\theta_B - \theta_S = \theta \Rightarrow \theta + \theta_S = \theta_B \Rightarrow \alpha = \theta_S$$

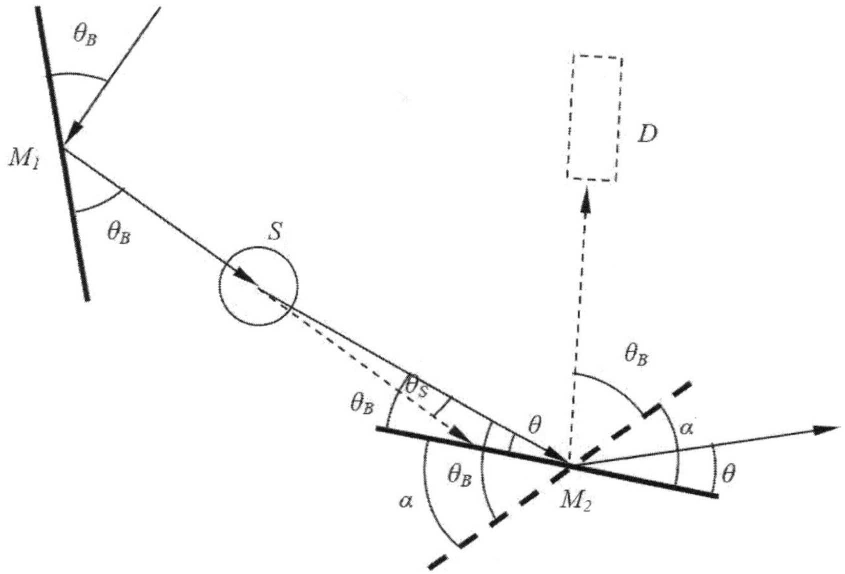

Figure 8. Three-axe antiparallel pattern.

To notice that this pattern is the most frequently used due to its dispersive feature this allows a better measurement of the interest neutrons.

3.2. The Method *I*(*Λ*) At Constant *Θ*

Another method is to make the observation only on a certain scattering direction 2θ = constant. This time the directly measurable quantity is the wavelength (or the energy) of the coherent radiation scattered by the sample. This method is based on λ determination for every detected neutron, by its time of flight measurement along a known distance. According to the De Broglie equation

$$\lambda = \frac{h}{mv} = \frac{ht}{mL}$$

where L is the distance between the neutron source and detector, h is the Planck constant, and m is the neutron mass. The neutrons with a wavelength λ will have a speed $v = h/m\lambda$ so that to cover an Ldistance, the time $t = L/v$ will be necessary.

Neutrons of different wavelengths will need different time to cover the same distance L. If the moment of neutron pulse emission coincides with the time scale origin of the neutron detector impulse analyzer, the detector impulse selection in accordance with the impulse appearance time (practically by neutron arrival time) allows the spectral distribution measurement of the neutron beam. The electronic devices performing this selection are "the time of flight analyzers." The analyzer has a number of channels in which the impulses are accumulated due to the neutrons arriving to the detector in the time intervals $t_0 + \tau$, $t_0 + 2\tau$, $t_0 + 3\tau$, and so on.

Here, t_0 represents a "delay time" of the impulse temporally analyzer system start-up, which could be varied with the used experimental pattern particularities, and τ is the channel temporal width, a quantity that can be varied in stages for a specific instrument. The measurements by the time of flight method can be made using both steady-state reactors and pulsed reactors as neutron sources.

In the case of steady-state reactors, the neutron beam leaving the reactor channel meets an obturator in its way which opens the way only for very short time intervals. In this manner, the detector "notices" a pulsed neutron beam with a determined time length, for a specific obturator–detector distance, by the neutron energetic spectrum. The detector is connected with a time of flight analyzer, which can be unleashed by the obturator opening. To the pulsed reactors, the obturator function is undertaken by the neutron source itself. The time of flight analyzer start-up command is given by the neutron source when the pulse is accomplished. In every SANS experiment, the main goal is to determine with a good precision and in a reasonably short time the Q dependence of the analyzed sample coherent elastic scattering differential cross section in a large as possible Q variation range. From the point of view of the Q range explored value extension, the two SANS investigation methods indicate the following possible performances:

$$\frac{Q_{\max}}{Q_{\min}} = \left[\frac{\theta_{\max}}{\theta_{\min}}\right]_{\lambda=\text{const.}} \quad \text{and respective} \quad \frac{Q_{\max}}{Q_{\min}} = \left[\frac{\lambda_{\min}}{\lambda_{\max}}\right]_{\theta=\text{const.}}$$

For any $Q_{\max} - Q_{\min}$, the interest interval for a certain sample, the first method seems to be more reasonable because the Q limitation apparently is dictated only by the measurement conditions, but in the case of the second method, the λ limitation is dictated by the spectral characteristics of the neutron source. If the neutron source is in steady state, then the first method utilization is certainly more advantageous because pulsing the neutron beam with a chopper leads to a usage of 10^{-4}–10^{-3} from the produced neutrons even if we consider the advantage of the second method to replace the sequential measurement of $I(Q)$ with the simultaneous one. The advantage of the first method toward the second in the case of utilization of a steady-state source is more evident if we estimate the consequences resulting in the obtained precisions and the experiment duration from the efficient utilization of the neutron beam emitted by the source. In the case of pulsed sources, the same criteria of neutron beam efficient utilization impose the application of the second working method. The approximate equivalence of the two methods, the first by the steady-state reactors and the second the pulsed sources, can be motivated that in present, at realized impulse fluxes and obtained pulses frequencies, the pulsed sources lead to getting middle fluxes comparable with the one obtained from the steady-state reactors of the same category. All these reasons justify the topics discussed in the next section, the presentation of SANS equipment used by the first method at the steady-state reactors or the second method used at the pulsed sources.

4. EXPERIMENTAL PATTERNS

4.1. The Method $I(Q)$ At Constant Λ

As it was discussed in Section 3, from among the optical patterns, the one with two crystals and the sample situated between them is much more appreciated. This pattern was used in many reactors, and a typical description of it is given in Fig. 9.

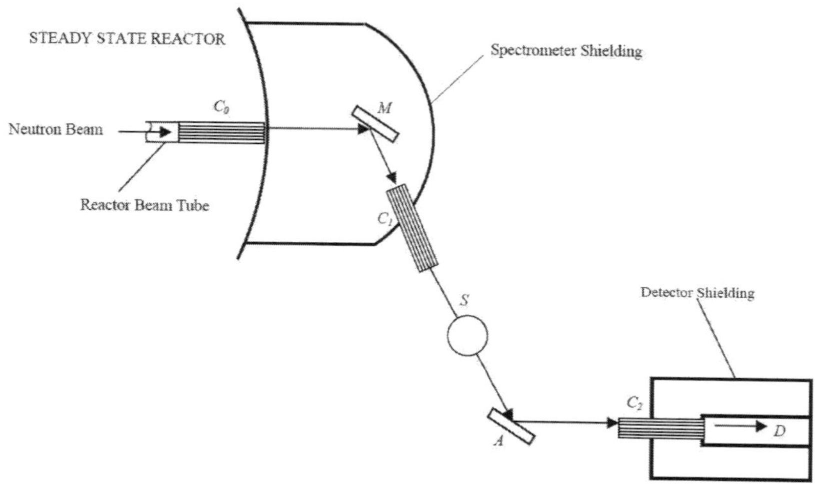

Figure 9. Three-axe spectrometer.

The monochromator group built by C_0 and C_1 collimators, and the monocrystal M sets up a monochromatic beam falling on the studied sample S. The analyzing group formed by the monocrystal$A$, the collimator C_2, and the detector D is initially tuned to detect the incident neutrons in the sample. To determine the neutron angular distribution produced by small angle scattering in the sample, we have to perform a combined scanning of the detector and the analyzer crystal around a vertical axis going through the center of the A crystal. Let us consider a small reactor providing a neutron flux of 1×10^{13} neutrons/cm^{2}s at 2 MW maximal power. From the white thermal neutron flux transmitted through the C_0 collimator, the M monochromator selects a very narrow band centered around the $\lambda = 1.15$ Å, with a wavelength resolution of $\Delta\lambda/\lambda = 4.5 \times 10^{-3}$. This beam is transmitted further through the C_1 collimator to the analyzed S sample and scattered at small angles (between 0 and 2 minutes). The A crystal collects the scattered beam and transmits it to the D detector through the C_2 collimator.

We consider that both crystals are copper monocrystals identically cut, with the reflecting plane orientation in the (002) direction and a mosaic divergence structure of 14 minutes. The C_0 and C_1 collimators have an angular horizontal divergence of 10 minutes, while that of the C_2 collimator is only of 5 minutes. The 2θ monochromator angular range is 50° but the scattering range of monochromator and analyzer is maximum 120°. The detection system is an ordinary BF_3 detector. Another type of neutron spectrometer is by using a cold source to allow neutron moderation at low temperatures, such as the VVR-SM reactor from Budapest Neutron Center (BNC). With an active core providing an average power of 10 MW, the neutron balance is improved by the mean of a beryllium reflector up to 10^{14} neutrons/cm^{2}s. The small angle scattering spectrometer described (Fig. 10) in the following is a conventional one. It uses a mechanical monochromator instead of a crystal monochromator. The mechanical selector with helical slits gives a wavelength resolution $\Delta\lambda/\lambda$ between 5 and 20% for a λ lying between 3.5 and 8 Å.

The neutron beam is geometrically built up by a 3 m-long vacuum collimator. The maximal beam dimension in the sample is 2 cm diameter. The distance from the sample to detector can vary from 1 to 5 m. The spectrometer is provided with an automatic sample changer controlled by a computer. The scattering vector range lies between 5×10^{-3} Å^{-1} < Q < 2 Å^{-1}. The scattered neutrons are detected by an XY positional detector with BF_3 having the dimension 64 × 64 cm^2.

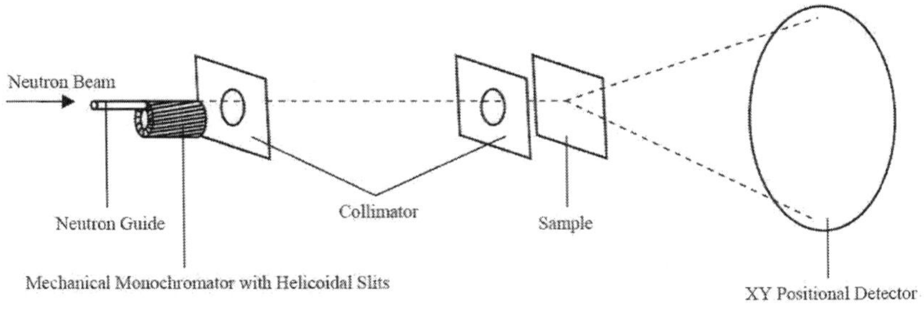

Figure 10. Spectrometer with mechanical monochromator.

Measurement instrument control and experimental data collection are performed by a computer.

4.2. The Method $I(\Lambda)$ At Constant Θ

This method is used mostly for the neutron pulsed sources. This type of spectrometer built on the time of flight principle is installed to the IBR-2 pulsed nuclear reactor at Joint Institute for Nuclear Research (JINR) in Dubna, Russian Federation. The name of the spectrometer is YuMo [7], given from its constructor, the scientist Yu. M. Ostanevich (Fig. 11). The reactor fuel is plutonium oxide, and uses liquid sodium as a coolant. The reactor reflector is composed of a static part and a rotational one with constructive and movement characteristics that determine the issue of fast neutron pulsed flux. The produced neutrons are moderated in water and through the horizontal beam tubes that surround the reactor active core in the experimental hall.

In the current working conditions, the reactor provides a maximum flux of thermal neutrons of 10^{16}neutrons/cm^{2}s per pulse at the moderator surface. The frequency of pulse repetition is 5 Hz.

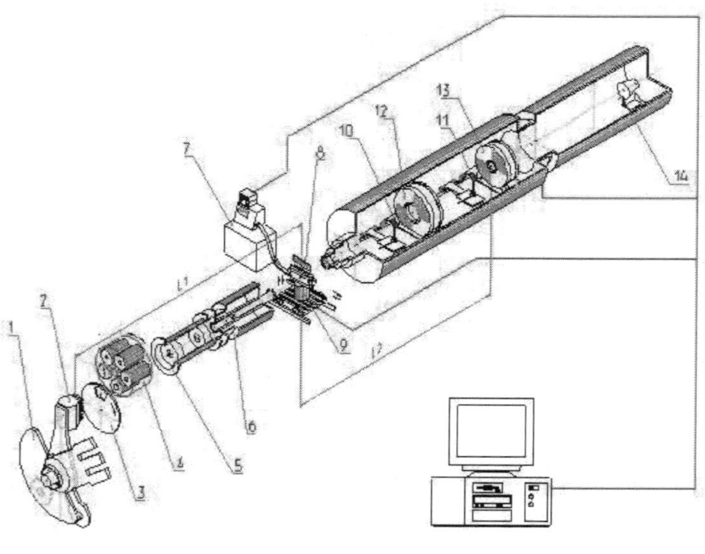

Figure 11. SANS spectrometer YuMo.1) two reflectors, 2) zone of reactor with moderator, 3) chopper, 4) first collimator (C1), 5) vacuum tube, 6) second collimator (C2), 7) thermostat, 8) samples table, 9) goniometer, 10)–11) vanadium standard, 12) ring-wire detector, 13) position-sensitive detector "Volga," 14) direct beam detector.

The chopper placed after the neutron moderator plays the main role to remove the neutrons from other reactor pulses and to assure that the neutrons from the interested pulse arrive at the detector. The neutron tube is made by steel pipes of different diameters and is kept under vacuum of about 10^{-2} Torr to avoid neutron absorption.

It is connected to the central vacuum system which is controlled from reactor control room and is provided with a mechanical beam stop that automatically closes the beam. A complete cycle of closing–opening the beam stop lasts about 6 minutes. The chopper is followed by a collimator with discrete variable aperture (C_1) that assures a geometrical pre-shape of the neutron beam and by the monitor whose signals are used for the normalization of the data obtained from different reactor pulses. The next collimator (C_2) assures the geometrical profile of the incident neutron beam in the sample and defines its dimensions. The collimator (C_1) is composed of four cylinders of different diameters (100 mm, 80 mm, 60 mm, and 40 mm), but with the same length which can be remote-controlled to be aligned in the neutron beam by a rotational movement in accordance with the experimental requirements of the beam geometrical pre-shape. The collimator (C_2) which limits the geometrical profile of the incident neutron beam in the sample has a remote-controlled variable opening (28 mm, 14 mm, and 7 mm). The holder with the collimation orifices of the (C_2) collimator are made from boron polyethylene having an insignificant scattering cross section in comparison with the absorption cross section. All collimators and the most important shielding elements are covered with a mixture of polyester and boron carbonate or boron acid. The sample holder having six slots is connected to a special remote-controlled mechanism of sample changing that allows each sample to be introduced in the neutron beam according to the experimental program. The sample is placed at 18.74 m from the moderator surface. The sample holder is provided with a cooling and warming system that allows adjusting the sample temperature between 20 to 150°C.

To record and analyze the scattered neutrons in accordance with their time of flight, the experimental equipment is provided with a measurement system having two detectors. The detectors are functionally identical but different by dimensions that assure their efficiency in observing the Q values at small and large ranges. Therefore, both detectors are provided with eight active ring detectors; one detector having the central opening of 80 mm while the second of 200 mm having outside diameters of 730 mm and 560 mm, respectively. Each detector can be settled in one of the eight positions having the sample detector distance fixed in the interval of 4.5 –12.5 m. The active part of each detector is divided into eight independent ring detectors by means of a concentric copper partition (the cathodes); at the center of each ring exists a tungsten wire of 25 μm thickness (the anode). This type of assembly provides eight independent ring detectors that offer more accuracy of the measurements by partial overlapping of the detection range of each ring with the range of the previous ring. The detector efficiency for neutrons with $\lambda = 1.8\text{Å}$ is about 70%.

The detector room is located after the samples and is made from a steel pipe of 1200 mm diameter and 12 m length having the inside surface covered with a cadmium layer of 0.5 mm thickness. Inside the room are mounted two rails for the movement of the electric-driven detectors trucks. The incident neutron beam is transmitted through the central hole of the detectors. At a distance of about 1.5 m from the reactor side, the detector is placed on a truck-driven mechanism in the vanadium standard (a vanadium metal sheet of about 0.3 mm thickness) that allows the introduction of the beam according to a program. This standard is used for the obtained experimental data calibration. The incident beam detector placed after the annular detectors measures the spectral distribution of the incident neutrons in the sample and forwards the normalization to this spectra of the scattered neutron spectra.

5. CONCLUSIONS

To summarize, we can say that small angle scattering is the collective name given to the techniques of SANS and SAXS scattering. In each of these techniques, radiation is elastically scattered by a sample, and the resulting scattering pattern is analyzed to provide information on the size, shape, and orientation of some components of the sample. They offer the possibility to analyze particles without disturbing their natural environment. The type of sample that can be studied by SAS, the sample environment that can be applied, the actual length scales that can be probed, and the information that can ultimately be obtained all depend on the nature of the radiation used. For example, SAXS cannot be used to study thick samples or samples requiring complex containers, while SANS can penetrate deeper in the condensed matter. SANS is produced by heterogeneities in matter. If these are randomly oriented, every atom pair contributes to the scattering of a sample. Inhomogeneities of sizes larger than atomic distances (10–1000 Å) produce scattering patterns with Q ranges $1/D < Q < 10/D$, if D is the dimension of the inhomogeneities.

A scattering experiment sees a scattering length density; in the case of X-rays, this is simply the electron density. The absolute density is not important but, in contrast, the difference between the particle and the surrounding medium.

The result of the experiment is the Fourier transformation of the contrast distribution. By comparing the experimental data with the theoretically calculated intensities or by a Fourier back transformation, we receive information about the contrast distribution, e.g., the mass density distribution of one particle. In the case of concentrated systems, we receive combined information about the single particle and the interaction between different particles.

In most cases, the sample and/or the sample environment are rather bulky.

Therefore, SANS instruments usually have to be large themselves in order to yield the desired resolution. Small SANS instruments can only serve a very limited number of applications. For reasons of intensity, a relatively large beam divergence, i.e., a beam cross section larger than the sample size is accepted as well as wavelength resolutions $\Delta\lambda/\lambda$ of up to about 20%.

SANS and SAXS techniques are complementary; however, they share several similarities. Perhaps the most important of these is the fact that, with minor adjustments to account for the different types of radiation, the same basic equations and "laws" (for example, those due to Guinier and Porod) can be used to analyze data from any of the two techniques.

This is a great advantage and one that has certainly eased the transition from one technique to another.

REFERENCES

1. Guinier A., Fournet G. Small-Angle Scattering of X-rays. New York: John Wiley &Sons Inc.; 1955. 268 p.

2. Bacon G. E. Neutron Diffraction. 3rd ed. Oxford: Claredon Press; 1975. 636 p.

3. Gerold V., Kostorz G. Small-angle scattering applications to materials science. J.Appl. Cryst. 1978; 11: 376–404.

4. Allen A. J., Windsor C. G., Rainey V., Pearson D., Double D. D., Alford N. M. A small angle neutron scattering study of cement porosities. J. Phys. D: Appl. Phys. 1982; 15:1817–1833.

5. Marshall W., Lovesey S. W. Theory of Thermal Neutron Scattering. 1st ed. Oxford:Claredon Press; 1971. 599 p.

6. Kostorz G., editor. Treatise on Materials Science and Technology. 1st ed. New York:Academic Press; 1979. 523 p.

7. Ostanevich Y. M. Time-of-flight small-angle scattering spectrometers on pulsed neutron

4

FORGE-HARDENED TIZR NULL-MATRIX ALLOY FOR NEUTRON SCATTERING UNDER EXTREME CONDITIONS

Takuo Okuchi[1,*], Akinori Hoshikawa[2] and Toru Ishigaki[2]

[1]Institute for Study of the Earth's Interior, Okayama University, 827 Yamada, Misasa, Tottori 682-0193, Japan

[2]Frontier Research Center for Applied Atomic Sciences, Ibaraki University, 162-1 Shirakata, Tokai, Naka, Ibaraki 319-1106, Japan

ABSTRACT

For neutron scattering research that is performed under extreme conditions, such as high static pressures, high-strength metals that are transparent to the neutron beam are required. The diffraction of the neutron beam by the metal, which follows Bragg's law, can be completely removed by alloying two metallic elements that have coherent scattering lengths with opposite signs. An alloy of Ti and Zr, which is known as a TiZr null-matrix alloy, is an ideal combination for such purposes. In this study, we increased the hardness of a TiZr null-matrix alloy via extensive mechanical deformation at high temperatures. We successfully used the resulting product in a high-pressure cell designed for high-static-pressure neutron scattering. This hardened TiZr null-matrix alloy may play a complementary role to normal TiZr alloy in future neutron scattering research under extreme conditions.

Keywords: neutron scattering; null-matrix alloy; high pressure; TiZr

1. INTRODUCTION

To conduct neutron scattering research under extreme conditions, such as neutron diffraction at high static pressures, metals that are transparent to neutrons and have a large mechanical strength are required for creating suitable sample environments [1]. The ideal metals for such applications should have small coherent and incoherent scattering lengths to reduce both the Bragg diffraction and background signals. The coherent scattering length of a metal can be minimized by preparing an alloy of two elements that have coherent scattering lengths with opposite signs. Such alloys are collectively called "null-matrix alloys", and the representatives are Mn (68 atom %) and Cu (32 atom %), which are ductile and soft [2], and Ti (68 atom %) and Zr (32 atom %), which are brittle and hard [3]. The latter $Ti_{68}Zr_{32}$ alloy (hereinafter referred to as TiZr) has been extensively used to fabricate metal gaskets for compressing samples in high-pressure environments, which are essential components of typical high-pressure cells designed for crystal structure analysis [4,5,6]. TiZr has the highest mechanical strength as a null-matrix alloy, which is roughly comparable to that of stainless steels [1]. TiZr also exhibits a moderate elongation of 6%–8% to be used in the aforementioned application [7].

In this study, by considering the unique properties of TiZr, we attempted to further enhance its mechanical properties while maintaining its null-scattering behavior via extensive mechanical deformation at high temperatures at ambient pressure. After preparing and analyzing the resulting product, we tested the applicability of the treated alloy to high-static-pressure experiments by using it in a high-pressure cell designed for neutron scattering.

2. EXPERIMENTAL METHODS

While a structural phase transition of TiZr null-matrix alloy was observed to occur as the effect of high static pressure [8], the effects of extensive deformation on its physical properties has not yet been studied. The metallographic textures and mechanical properties of Ti–Zr two-component α-alloy system can be modified and improved by subjecting them to plastic deformations (via mechanical forging) [9]. Therefore, here we prepared a cast billet of TiZr and applied extensive plastic deformation for it at ambient pressure [10,11]. A total of 85 kg of the TiZr alloy was cast and forged as a custom-made commercial order to Daido Bunseki Research, Inc. (Nagoya, Japan). In detail, the weighted Ti- and Zr-containing precursors were first melted together in a plasma progressive-casting furnace, where an electrode for the subsequent vacuum arc remelting (VAR) procedure was prepared. The electrode was then subjected to the VAR procedure, which produced a homogenized ingot after the remelting and recooling. The appearance of the ingot is shown inFigure 1, while the composition is shown in Table 1.

Figure 1. Photograph of the VAR-fabricated TiZr ingot, which was 240 mm in diameter, 384 mm in length, and 85.4 kg in weight.

Table 1. Chemical composition of the vacuum arc remelting (VAR)-fabricated TiZr ingot.

Composition wt. %							
Zr	C	Al	B	O	N	Fe	Ti *
47.1	0.008	0.01	<0.01	0.053	0.004	0.03	52.8

* Ti as the balance.

The VAR-fabricated TiZr ingot was then polished to remove any macroscopic cracks in the surface. An evacuated steel sheath was used to subject the billet to vacuum conditions during the forging process, which is an effective method for preventing the oxidation of the ingot. The ingot-containing steel sheath was heated to approximately 1100 °C and then forged immediately. The diameter of the ingot was then reduced to approximately 100 mm as the temperature decreased to around 700 °C.

The thinned ingot (Figure 2) was then cut, re-sealed in another steel sheath, and forged again into a number of cylinders with diameters of 20–30 mm. This process was conducted at temperatures within the β-phase field, and the final products were air-cooled into the α-phase [12]. The final products had a forging ratio (F_0/F) of more than 60, or $e = \ln(F_0/F) = 4–5$, where F_0 and F are the initial and final cross-sectional areas, respectively. These values were sufficient to produce the desired metallographic texture. Hereinafter, we will refer to the final products as "f-TiZr", where the "f" indicates that the resulting products were both "forged" and "finely textured".

The metallographic textures and elemental concentration maps of the f-TiZr alloy were obtained with a field-emission scanning electron microscopy (FE-SEM; JEOL JSM-7001F, Akishima, Tokyo, Japan) setup that was capable of performing energy dispersive X-ray spectroscopy (EDS). The micro-Vickers hardness of the alloy was measured using a micro hardness tester with 300 g to 500 g of the applied force (Mitsutoyo HM-221, Kawasaki, Japan). Powder X-ray diffraction (XRD; Rigaku Smartlab, Akishima, Tokyo, Japan) was used to analyze the crystal structure, while neutron powder diffraction (NPD) was used to evaluate the coherent neutron scattering. The XRD patterns were

Figure 2. Photograph of the forged ingot that is approximately 100 mm in diameter, which was mechanically thinned from the VAR-fabricated TiZr ingot that was 240 mm in diameter.

obtained by using the para-focusing method with Cu K$\alpha_1$radiation operated at 40 kV and 30 mA. The NPD patterns were obtained with an iMATERIA pulsed-neutron powder diffractometer at the Materials and Life Science Experimental Facility of the Japan Proton Accelerator Research Complex (J-PARC; Tokai, Japan). The neutron diffractometer covered a wide range of*d*-values (0.26–40 Å) in the double-frame operating mode [13].

We expected that the toughness and elongation of the f-TiZr would not be sufficient for certain applications that involve extensive deformations of the metal. Therefore, some of the samples were cut from the forged TiZr ingot with 100 mm in diameter (Figure 2) and were annealed and furnace-cooled at 660 °C for 6 h in an Ar atmosphere using a gas convertible vacuum furnace (FUA112DB, Advantec Toyo Kaisya, Ltd., Tokyo, Japan), which is an effective method for increasing the toughness and elongation of the alloy while reducing its hardness [9].

3. RESULTS AND DISCUSSION

Figure 3 shows typical cross-sectional SEM images of the VAR-fabricated TiZr ingot before it was forged. The back-scattered electron images show a contrast on the averaged atomic number, where the compositional heterogeneity is heavily emphasized. Upon cooling the ingot after its solidification, the temperature passes through the phase boundary between the body-centered cubic *Im*3*m* (β) phase and the hexagonal close-packed $P6_3/mmc$ (α) phase, causing the recrystallization of α-platelets. This in turn creates the Widmanstätten patterns, which are similar to that observed in some furnace-cooled Ti–Zr alloys with different compositions [9]. The typical lamellar width in the VAR-fabricated TiZr ingot was 5–50 μm. TiZr alloys with such textures have been used as null-matrix alloys for neutron scattering.

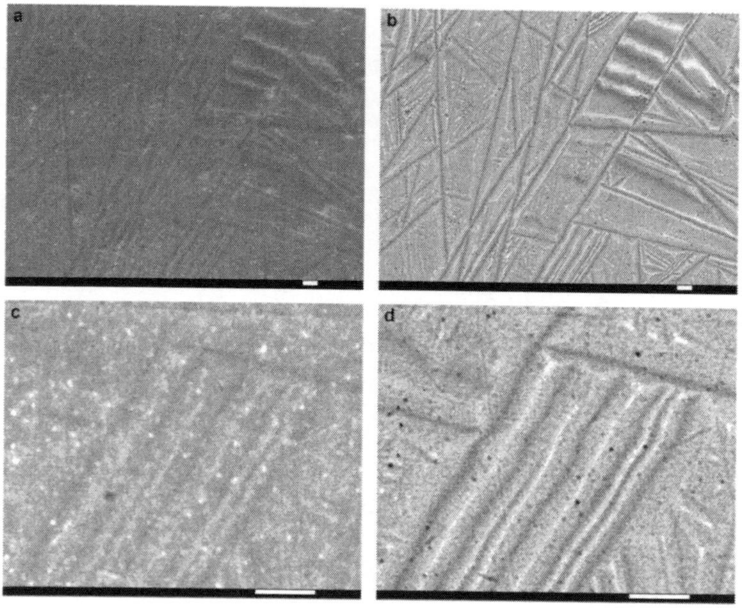

Figure 3. Cross-sectional (**a**) secondary-electron and (**b**) back-scattered-electron SEM images of the VAR-fabricated TiZr ingot before it was subjected to the mechanical forging. (**c**,**d**) Magnified images of areas in (**a**) and (**b**), respectively (magnification = 4×). All scale bars indicate a length of 10 μm.

Figure 4 shows SEM images and elemental concentration maps of a polished cross section of the f-TiZr alloy, which was cut from a forged cylinder that was 20 mm in diameter. The very fine metallographic textures shown in these images should be effective to increase the hardness and strength of the metal [14], which has been repeatedly confirmed by previous studies, including those of Ti-based alloy systems [15,16,17,18]. This is also shown in our measurements of micro-Vickers hardness of the f-TiZr alloy at H_v = 339, which is significantly higher than that of non-forged $Ti_{60}Zr_{40}$ or $Ti_{70}Zr_{30}$ alloy with H_v = 230~280 [19]. The typical width of the α-platelets in the f-TiZr alloy was 0.1–0.5 μm, which is approximately two orders of magnitude finer than that observed in the VAR-fabricated TiZr ingot (Figure 3).

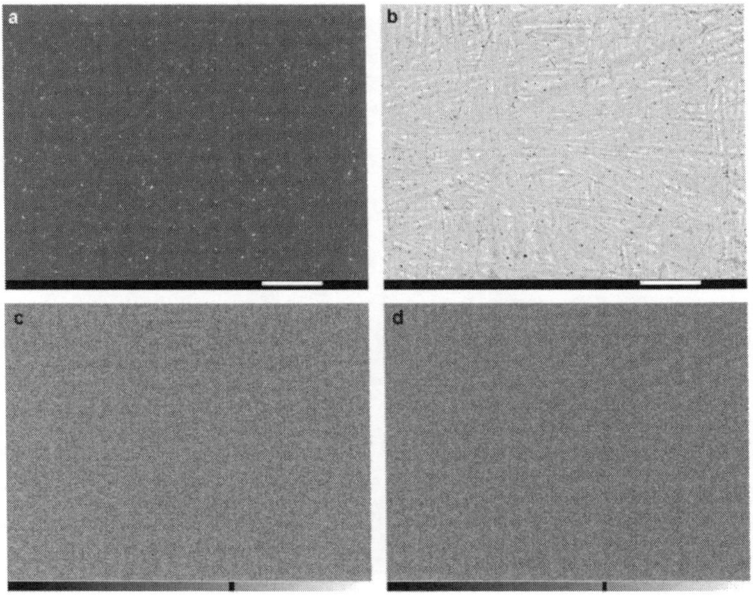

Figure 4. Cross-sectional (**a**) secondary-electron and (**b**) back-scattered-electron SEM images of the f-TiZr alloy along with the scale bars with 10 μm in length. (**c**) Ti and (**d**) Zr concentration maps of the f-TiZr alloy obtained by EDS along with intensity scale bars (below each image where the left-end color shows no intensity) and averaged intensity (black line in each bar). All these images have identical spacial scale to that of Figure 3c,d.

Figure 5 shows a typical XRD pattern of the f-TiZr alloy. All of the reflections were successfully indexed to those of the α-phase, as indicated by the indices of each peak. A significant broadening of these reflections is consistent with a large internal strain within the α-platelets; this strain is essential for extensively hardening α-Ti alloys [9,19,20]. Figure 6 shows the NPD patterns of the f-TiZr alloy. The patterns show that there are no diffraction peaks from the f-TiZr for the range of *d*-values measured. Therefore, we have confirmed that the forging process performed on the TiZr alloy has not degraded its null-scattering properties.

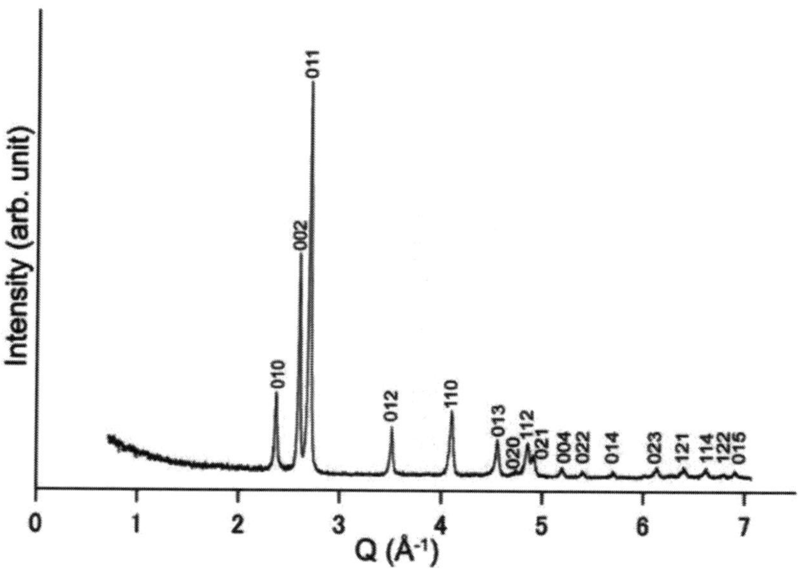

Figure 5. X-ray diffraction (XRD) pattern of the f-TiZr alloy. The Miller indices of the α-phase are shown next to the corresponding reflections.

Figure 7 shows SEM images and elemental concentration maps of the annealed f-TiZr alloy. The width of the lamellae is smaller than 10 μm, and thus, the lamellae have not returned to the original dimensions of the VAR-fabricated TiZr ingot. On the other hand, its well-ordered metallographic texture is similar to that of the VAR-fabricated TiZr ingot or a furnace-cooled alloy [9]. The measured micro-Vickers hardness of the annealed f-TiZr was H_v = 279, which was reduced from the original hardness of the f-TiZr (H_v = 339).

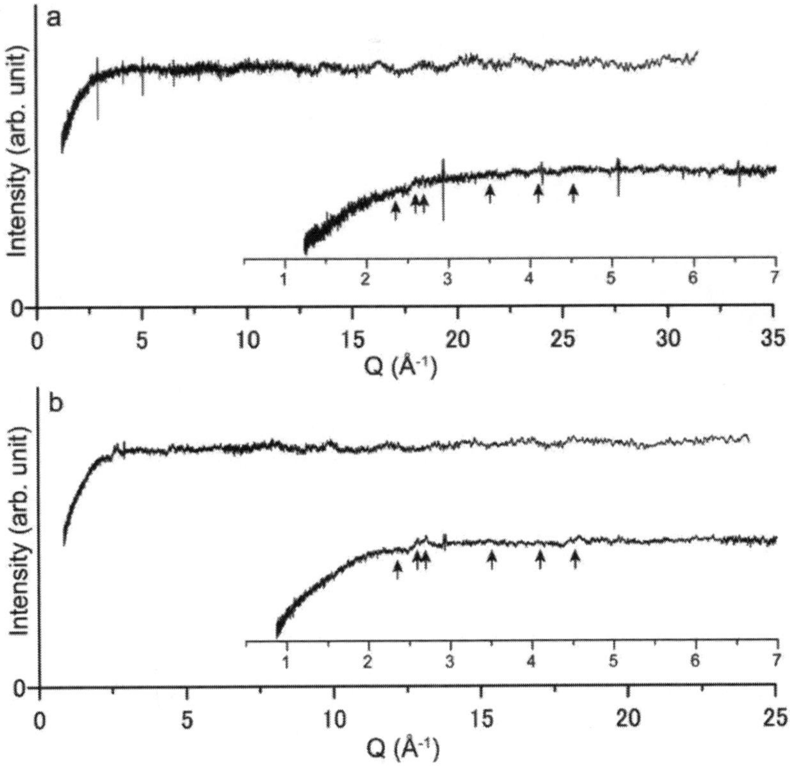

Figure 6. Neutron powder diffraction (NPD) patterns of the f-TiZr alloy: (**a**) NPD pattern from the high-resolution bank ($\Delta d/d \approx 0.16\%$) and (**b**) that from the special environment bank ($\Delta d/d \approx 0.5\%$). The intensities of both profiles initially increase with increasing Q, which is due to inelastic scattering. The inset profiles show the lowest Q regime which have the same vertical scale with the main figures. Arrows under both of the inset profiles indicate the positions of representative Bragg reflections that are observable when TiZr has a non-ideal composition; their indices include 010, 002, 011, 012, 110, and 013 for the α phase (seeFigure 5). These Bragg reflections could be more evident for the TiZr used in the previous research (*i.e.*, Figure 2.5 in Reference [1]). A few spurious, narrow peaks involving both negative and positive anomalies present in the profiles (at 2.93, 4.16 and 5.11 Å^{-1}), which were not completely removed by the background-subtraction procedures. These are caused by the diffraction of the sample holder, which was composed of *V*.

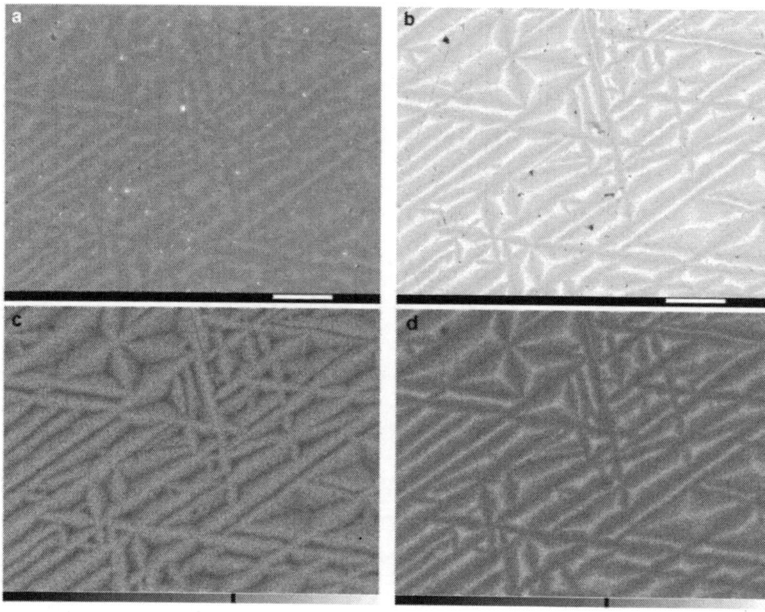

Figure 7. Cross-sectional (**a**) secondary-electron and (**b**) back-scattered-electron SEM images of the annealed f-TiZr alloy along with the scale bars with 10 μm in length. (**c**) Ti and (**d**) Zr concentration maps of the annealed f-TiZr alloy obtained by EDS along with intensity scale bars (below each image where the left-end color shows no intensity) and averaged intensity (black line in each bar). All these images have identical spacial scale to that of Figure 3c,d.

4. APPLICATION OF THE F-TIZR ALLOY

A series of experiments were conducted by fabricating deformable metallic gaskets from the f-TiZr alloy, and static pressures greater than 15 GPa were successfully generated without the gaskets fracturing. We used a newly designed high-pressure cell that was optimized for neutron-scattering measurements with strong pulsed-neutron sources [21,22]. In brief, two anvils composed of single-crystal moissanite and sintered polycrystalline diamond with 5 mm culets were used to compress a powdered $Mg(OD)_2$ sample, which was confined within the f-TiZr gasket (initial thickness = 1 mm) and then compressed between the two anvils to 9 GPa. The total sample volume compressed to this high pressure was approximately 3 mm^3, which was sufficient for performing a precise structure refinement via NPD at J-PARC. The gasket was recovered after this compression experiment, and it is shown in Figure 8. Similar to the other successful cases, the gasket has not cracked, even after the extensive deformation.

The annealed f-TiZr alloy was also applied to high-pressure NPD measurements with various types of anvil geometry, the results of which were partially published elsewhere [23].

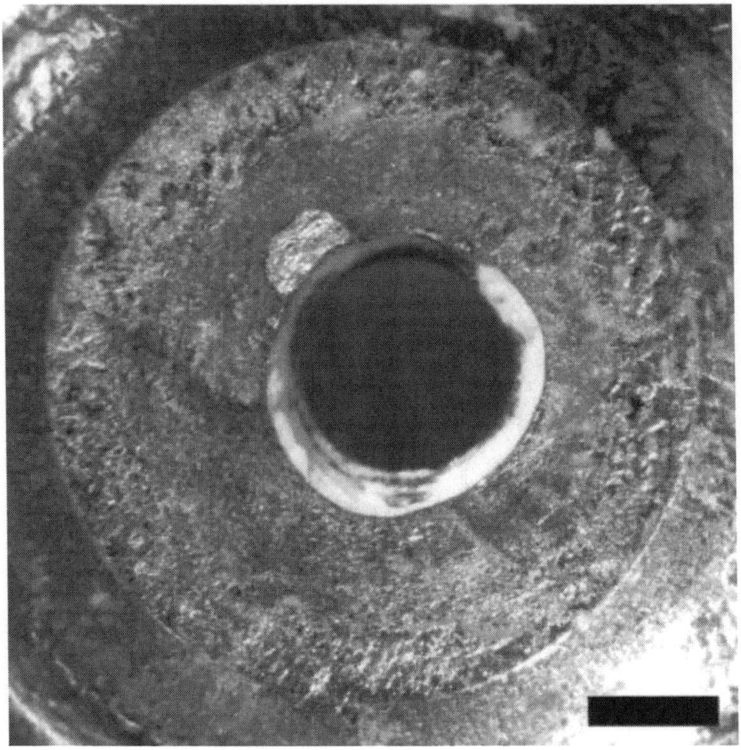

Figure 8. The recovered f-TiZr gasket, which experienced extensive deformation in the high-pressure cell. It was used to compress a powdered $Mg(OD)_2$ sample to 9 GPa. The scale bar indicates a length of 1 mm.

5. CONCLUSIONS

By applying extensive mechanical deformation at high temperatures to a TiZr alloy, the metallographic textures of the alloy became much finer, with the lamellae width decreasing by two orders of magnitude. The mechanical hardness of the TiZr was increased because of this change in the metallographic texture, which suggests that its strength was also increased, as previously observed in the general Ti–Zr two-component alloy system [9]. This f-TiZr alloy was successfully applied to experiments involving the generation of high static pressures, which requires the extensive deformation of the hard metal. In addition, the f-TiZr alloy was also used to fabricate a thin-walled micro-sample container, which enabled us to obtain a background-free NPD pattern with a pulsed-neutron source [24,25]. The successful application of the f-TiZr alloy has demonstrated that forged TiZr products may play a complementary role in future neutron scattering research.

ACKNOWLEDGMENTS

The authors wish to thank T. Iikubo at Daido Bunseki Research Co, Ltd. and T. Matsumoto at the University of Tokyo for providing technical assistance. The machine time for the pulsed-neutron diffraction experiments was provided through the J-PARC MLF user program (No. 2013B0279). N. Tomioka at Okayama University is acknowledged for helping these experiments. Four anonymous referees are acknowledged for their comments to improve the manuscript. This work was supported in part by JSPS Grants-in-Aid for Scientific Research (No. 19GS0205, 26287135 and 15K13593).

AUTHOR CONTRIBUTIONS

T.O. wrote this manuscript and performed the experiments to obtain all of the results except for the neutron diffraction patterns. T.O., A.H., and T.I. performed the experiments to obtain the neutron diffraction patterns. All authors contributed to the interpretation of the experimental data.

CONFLICTS OF INTEREST

The authors declare no conflicts of interest.

REFERENCES

1. Klotz, S. Techniques in High Pressure Neutron Scattering; CRC Press: Boca Raton, FL, USA, 2012.

2. Smith, J.H.; Vance, E.R.; Wheeler, D.A. A null-matrix alloy for neutron diffraction. J. Phys. E Sci. Instrum. 1968, 1, 945–947.

3. Sidhu, S.S.; Heaton, L.; Zauberis, D.D.; Campos, F.P. Neutron diffraction study of titanium-zirconium system. J. Appl. Phys. 1956, 27, 1040–1042.

4. Besson, J.M.; Nelmes, R.J.; Hamel, G.; Loveday, J.S.; Weill, G.; Hull, S. Neutron powder diffraction above 10 GPa. Phys. B Condens. Matter 1992, 180–181, 907–910.

5. Klotz, S.; Besson, J.M.; Hamel, G.; Nelmes, R.J.; Loveday, J.S.; Marshall, W.G.; Wilson, R.M. Neutron powder diffraction at pressures beyond 25 GPa. Appl. Phys. Lett. 1995, 66, 1735–1737.

6. Xu, J.; Ding, Y.; Jacobsen, S.D.; Mao, H.K.; Hemley, R.J.; Zhang, J.; Qian, J.; Pantea, C.; Vogel, S.C.; Williams, D.J.; et al. Powder neutron diffraction of wustite (Fe0.93O) to 12 GPa using large moissanite anvils. High Press. Res. 2004, 24, 247–253.

7. Williams, D.N.; Wood, R.A.; Jaffee, R.I.; Ogden, H.R. The effects of zirconium in titanium-base alloys.J. Less Common Met. 1964, 6, 219–225.

8. Zeidler, A.; Guthrie, M.; Salmon, P.S. Pressure-dependent structure of the null-scattering alloy Ti0.676Zr0.324. High Press. Res. 2005, 35, 239–246.

9. Imgram, A.G.; Williams, D.N.; Ogden, H.R. Tensile properties of binary titanium-zirconium and titanium-hafnium alloys. J. Less Common Met. 1962, 4, 217–225.

10. Meyers, M.A.; Mishra, A.; Benson, D.J. Mechanical properties of nanocrystalline materials. Prog. Mater. Sci. 2005, 51, 427–556.

11. Sakai, T.; Belyakov, A.; Kaibyshev, R.; Miura, H.; Jonas, J.J. Dynamic and post-dynamic recrystallization under hot, cold and severe plastic deformation conditions. Prog. Mater. Sci. 2014, 60, 130–207.

12. Blacktop, J.; Crangle, J.; Argent, B.B. The $\alpha \rightarrow \beta$ transformation in the Ti–Zr system and the influence of additions of up to 50 at. % Hf. J. Less Common Met. 1985, 109, 375–380.

13. Ishigaki, T.; Hoshikawa, A.; Yonemura, M.; Morishima, T.; Kamiyama, T.; Oishi, R.; Aizawa, K.; Sakuma, T.; Tomota, Y.; Arai, M.; et al. IBARAKI materials design diffractometer (iMATERIA)—Versatile neutron diffractometer at J-PARC. Nucl. Instrum. Meth. A 2009, 600, 189–191.

14. Hall, E.O. Yield Point Phenomena in Metals and Alloys; Plenum Press: New York, NY, USA, 1970; p. 316.

15. Herebtsov, S.; Lojkowski, W.; Mazurb, A.; Salishcheva, G. Structure and properties of hydrostatically extruded commercially pure titanium. Mater. Sci. Eng. A 2010, 527, 5596–5603.

16. Salishchev, G.A.; Galeev, R.M.; Malysheva, S.P.; Zherebtsov, S.V.; Mironov, S.Y.; Valiakhmetov, O.R.; Ivanisenko, E.I. Formation of submicrocrystalline structure in titanium and titanium alloys and their mechanical properties. Met. Sci. Heat Treat. 2006, 48, 63–69.

17. Valiev, R.Z.; Alexandrov, I.V. Paradox of strength and ductility in metals processed by severe plastic deformation. J. Mater. Res. 2002, 17, 5–8.

18. Kobayashi, Y.; Tanaka, Y.; Matsuoka, K.; Kinoshita, K.; Miyamoto, Y.; Murata, H. Effect of forging ratio and grain size on tensile and fatigue strength of pure titanium forgings. J. Soc. Mater. Sci. Jpn. 2005, 54, 66–72.

19. Kobayashi, E.; Matsumoto, S.; Doi, H.; Yoneyama, T.; Hamanaka, H. Mechanical properties of the binary titanium-zirconium alloys and their potential for biomedical materials. J. Biomed. Mater. Res.1995, 29, 943–950.

20. Dobromyslov, A.V.; Taluts, N.I. Structure investigation of quenched and tempered alloys of the Zr–Ti system. Phys. Met. Metallogr. 1987, 63, 114–120.

21. Okuchi, T.; Sasaki, S.; Ohno, Y.; Abe, J.; Arima, H.; Osakabe, T.; Hattori, T.; Sano-Furukawa, A.; Komatsu, K.; Kagi, H.; et al. Neutron powder diffraction of small-volume samples at high pressure using compact opposed-anvil cells and focused beam. J. Phys. Conf. Ser. 2012.

22. Okuchi, T.; Yoshida, M.; Ohno, Y.; Tomioka, N.; Purevjav, N.; Osakabe, T.; Harjo, S.; Abe, J.; Aizawa, K.; Sasaki, S. Pulsed neutron powder diffraction at high pressure by a capacity-increased sapphire anvil cell. High Press. Res. 2013, 33, 777–786.

23. Okuchi, T.; Tomioka, N.; Purevjav, N.; Abe, J.; Harujo, S.; Gong, W. Structure refinement of sub-cubic-mm volume sample at high pressures by pulsed neutron powder diffraction: Application to brucite in an opposed anvil cell. High Press. Res. 2014, 34, 273–280.

24. Purevjav, N.; Okuchi, T.; Tomioka, N.; Abe, J.; Harjo, S. Hydrogen site analysis of hydrous ringwoodite in mantle transition zone by pulsed neutron diffraction. Geophys. Res. Lett. 2014, 41, 6718–6724.

25. Tomioka, N.; Okuchi, T.; Purevjav, N.; Abe, J.; Harjo, S. Hydrogen sites in the dense hydrous magnesian silicate phase E: A pulsed-neutron powder diffraction study. Phys. Chem. Miner. in press.

5

INVESTIGATING THE DEFECT STRUCTURES IN TRANSPARENT CONDUCTING OXIDES USING X-RAY AND NEUTRON SCATTERING TECHNIQUES

Gabriela B. González

Department of Physics, DePaul University, 2219 N. Kenmore Avenue, Chicago, IL 60614, USA

ABSTRACT

Transparent conducting oxide (TCO) materials are implemented into a wide variety of commercial devices because they possess a unique combination of high optical transparency and high electrical conductivity. Created during the processing of the TCOs, defects within the atomic-scale structure are responsible for their desirable optical and electrical properties. Therefore, studying the defect structure is essential to a better understanding of the behavior of transparent conductors. X-ray and neutron scattering techniques are powerful tools to investigate the atomic lattice structural defects in these materials. This review paper presents some of the current developments in the study of structural defects in n-type TCOs using x-ray diffraction (XRD), neutron diffraction, extended x-ray absorption fine structure (EXAFS), pair distribution functions (PDFs), and x-ray fluorescence (XRF).

Keywords: transparent conducting oxide; indium-tin oxide; tin oxide; zinc oxide; defect; x-ray diffraction; neutron diffraction; EXAFS; XRF; PDF

1. INTRODUCTION

In most materials, high transparency in the visible region and high electrical conductivity are mutually exclusive properties. However, thin films of transparent conducting oxide (TCO) materials are approximately 90% transparent in the visible region and exhibit electrical conductivity of more than 1000 S/cm [1]. This unique combination of these two properties is very attractive for a wide variety of commercial applications, where TCO materials act as electrodes in optoelectronics devices, such as flat-panel displays and smart windows [2]. Tin oxide was the first commercialized TCO [1]. Currently, commercially-produced TCOs are n-type semiconductors, which include indium oxide, tin oxide, zinc oxide, gallium oxide, and cadmium oxide, that are appropriately doped to improve their electrical behavior while maintaining their transparency. Tin oxide and indium-tin oxide (ITO) are the most widely used TCOs. One advantage of ITO over both pure In_2O_3 and SnO_2 is that ITO films etch more easily and exhibit ideal chemical stability for various device applications. For flat-panel displays the TCO of choice is ITO due to its superior properties and industrial performance.

The transparency of a TCO is due to the wide band gap of the material, which is at least 3.1 eV, corresponding to the highest energy of visible light. The electrical conductivity of a TCO results from the presence of atomic-scale defects, which for n-type semiconductors, bring in donors that increase the electron population. Because the intrinsic population of atomic-scale defects in these oxides is typically very small, the optimization of TCO performance consists of introducing more desirable defects during their synthesis. A post-anneal step in a reducing atmosphere is usually performed to increase the electron population.

The specific types of atomic-scale defects for a given TCO depend on the specific oxide and its synthesis. They may include oxygen vacancies, cation vacancies, oxygen interstitials, cation interstitials, impurity dopants, cation and anion anti-sites, and defect complexes, which consist of a combination of two or more point defects. It is conjectured that more than one type of defect may be present in a material, but the more abundant, non-neutral defects have the greatest effect on the overall conductivity. The charge of these defects is very important, since the overall charge compensation dictates the amount of free donors. In several TCOs with extended dopant concentrations, the dopant population is much larger than the equilibrium limit. High conductivity is expected for these largely doped TCOs, but it is not always observed due to the possible presence of neutral defect complexes which can store the dopants so that they do not contribute to the free electron population [3,4].

It is evident that determining the defect chemistry of TCOs is essential to the understanding of the behavior of these materials and that the optimization of their synthesis would enable an even wider range of commercial applications. Theoretical defect models for TCOs have been developed, but many times the theoretical predictions found in the literature are conflicting [5,6,7,8,9] and lack experimental confirmation. Electrical measurements can provide insight on potential defect models [4]. X-ray and neutron scattering are powerful experimental tools to directly investigate the atomic lattice structural defects in materials. A combination of electrical and scattering measurements can offer the much needed experimental evidence for defect models. This review paper presents some of the current developments in the study of defects in indium-based, tin-based, and zinc-based n-type TCOs using x-ray and neutron scattering techniques.

2. DESCRIPTION OF SCATTERING TECHNIQUES

2.1. X-ray and Neutron Diffraction

X-ray and neutron diffraction are techniques that give direct information about the structural arrangement of atoms in materials. X-rays interact with the electron cloud surrounding an atom. When the interaction is elastic, the scattered signal contains quantitative information about the location of atoms as well as the type of atoms present. The x-ray scattering factor, $f(E)$, is expressed as:

$$f(E) = f_0 + f'(E) + if''(E) \quad (1)$$

where f0 is the energy-independent form factor, which is directly related to the number of electrons; f' and f'' are the anomalous dispersion corrections, which have major contributions only at certain energies.

Because the scattering factor at most x-ray wavelengths is dominated by the f0 factor, the x-ray signal is approximately proportional to the number of electrons, and high *Z* elements scatter more strongly than light elements. Therefore, if two elements present in a material are next to each other in the periodic table, their x-ray scattering powers will be nearly indistinguishable at all wavelengths except at the x-ray absorption edges when anomalous scattering effects become important. One example of a TCO where this problem is encountered is indium-tin oxide, where indium and tin have 49 and 50 electrons, respectively, making them almost indistinguishable at most x-ray wavelengths. If a tin atom substitutes for an indium atom on a given site, this substitutional defect could not be quantified with a conventional, lab-based x-ray source. A similar case is antimony-doped tin oxide because antimony has 51 electrons. For these types of materials, an x-ray synchrotron source is necessary so that the experimenter can tune the wavelength to near the absorption edges where the anomalous scattering contributions are enhanced to maximize the contrast between the elements. Alternatively, neutrons instead of x-ray photons can be used to diffract from the samples.

Neutron diffraction follows the same basic principles as x-ray diffraction (XRD) and the neutron scattering length is equivalent to the x-ray scattering factor. One primary difference is that the neutron scattering length is independent of the number of electrons, since the scattering centers are the neutrons of the atoms. This is especially advantageous in the study of light elements such as oxygen, which suffers from a weak x-ray signal but exhibits a strong neutron signal (scattering length is 5.803(4) fm). For TCOs where oxygen vacancies and oxygen interstitials are important defects, neutron diffraction is a highly valuable technique. Another advantage of neutron diffraction is that most elements with similar number of electrons can be distinguished as in the case of ITO and Sb-doped SnO_2. The neutron scattering lengths are 4.0652(20)–i9.9539(4) fm for indium, 6.2257(19) fm for tin, and 5.57(3) fm for antimony.

X-ray and neutron diffraction data also contain information about the microstructural defects of crystalline materials. Nanometer-sized grains and lattice strain increase the width of the diffraction peaks. Careful analysis of the peak shape obtained from high-resolution instruments can provide an estimate of the average crystallite size and strain. Crystalline materials produce well defined Bragg peaks that result from the interaction of x-rays with stacks of regular atomic planes. Analysis of the position and intensity of Bragg peaks is used to solve the average, long-range structure of the material. For the case of amorphous materials, where a long-range periodic arrangement of atoms does not exist, the diffraction signal does not contain sharp, strong peaks but rather broad, weak peaks. In those cases, other scattering techniques may be more appropriate to learn about the short-range order of atoms. A brief description of some of those scattering techniques is given below.

2.2. Extended X-ray Absorption Fine Structure (EXAFS) and X-ray Absorption near Edge Spectroscopy (XANES)

EXAFS is an x-ray scattering technique that can provide valuable information about the local structure of both crystalline and amorphous materials. When an x-ray beam traverses a sample of finite thickness, some of the x-ray signal is absorbed by the sample. The absorption coefficient of the sample depends on the energy of the incident x-ray beam and the sample composition. If the energy of the x-ray photon is high enough to excite electrons from inner shells, more electrons become available for photoemission, and the absorption coefficient increases sharply. The energies where discontinuous increases in the absorption coefficient of the sample occur are called absorption edges and are labeled *K, L, M,* and so on, depending on which shell energy of the absorber electron the edge matches.

The high-energy sides of an absorption edge have oscillations for up to 2000 eV. This is referred to as the extended x-ray absorption fine structure (EXAFS) region. The photoelectrons that are excited from inner shells interfere with the scattered wave produced by the atoms neighboring the absorbing atom. The interference can be constructive or destructive producing peaks and valleys in the EXAFS region. The EXAFS features therefore contain quantitative information about the neighboring atoms surrounding the absorbing element. In particular, coordination number, distance, and Debye-Waller factors can be obtained. The x-ray absorption near edge spectroscopy (XANES) region includes only about 50 eV immediately after the absorption edge. From this region, chemical information, such as coordination number and valence state of the absorbing element, is obtained. For EXAFS and XANES measurements an x-ray synchrotron source is typically required to scan through the x-ray absorption edge of the element of interest. High statistics, high energy resolution, and a stable monochromatized x-ray beam are essential for successful experiments.

2.3. Atomic Pair Distribution Functions (PDFs)

Atomic PDFs can be obtained for both crystalline and amorphous materials since no long-range order or periodicity is required. Both the Bragg-like peaks and the diffuse scattering component of the x-ray diffraction data can be analyzed. Unlike diffraction, PDFs peaks do not reflect the periodicity found in atomic planes but rather arise from atomic pairs, coordination distances and spheres. A high-energy x-ray or neutron source is typically needed to collect data with a wide range of scattering angles. High-intensity flux and efficient detectors are also required to obtain signals with good statistics. The analysis of the data involves extracting the coherent component of the intensity from experimental patterns using careful corrections. The results consist of radial distributions similar to those obtained from EXAFS, but in this case the information is not element specific. A PDF peak represents a pair of atoms separated by a distance *r*. Similar to the x-ray diffraction experiments where anomalous contributions are enhanced, it is possible to collect data at resonant wavelengths and perform differential analysis to highlight correlations between particular atomic pairs. More detailed explanations on this technique can be found in the literature [10,11,12,13].

2.4. X-ray Fluorescence (XRF)

As highlighted before, the presence of impurities and dopants in TCOs are important since they can enhance or diminish the electrical properties of the material. The detection and quantification of elements can be obtained via chemical analysis. X-ray fluorescence (XRF) is a non-destructive technique that is commonly used for elemental analysis, and it can detect concentrations as low as a few part per million. A source of energetic photons (x-ray or gamma rays) is required to excite inner electrons from *K* and *L* shells. When electrons from higher shells fill in the inner shells, characteristic secondary radiation is emitted in the form of x-ray photons. This radiation is referred to as x-ray fluorescence. The sample spectrum contains fluorescent lines specific to the elements present in the sample. Analysis of the peak intensities provides quantification of each element. Typical commercial XRF instruments can detect elements as light as magnesium; lighter elements are not easily measured due to attenuation of their signals. This technique, unlike the previous ones, does not provide information about the atomic arrangement of the elements in the sample, but it can be applied to both crystalline and amorphous samples to independently quantify elemental compositions.

3. RESULTS ON DEFECTS IN TCOS

In this manuscript point defects will be expressed using Kröger-Vink notation. The superscripts denote the charge of the defects: a neutral defect is represented by an (^{x}), a negative defect with ('), and a positive defect with a (•) superscript. The subscripts describe the site where the defect is located. For example, OxOrepresents a neutral oxygen atom on an oxygen site; V··O represents a vacancy on an oxygen site having an effective positive charge of +2; and e′ represents an electron with one negative charge.

3.1. Bixbyite In_2O_3

The In_2O_3 structure was studied using x-ray diffraction on single crystals at two wavelengths [14]. Indium oxide crystallizes in the cubic bixbyite structure, also known as C-type rare-earth sesquioxide with the space group Ia3 (number 206). Its lattice parameter is 10.117(1) Å, and a unit cell has 32 cations and 48 anions. The bixbyite structure is derived from a 2 × 2 × 2 fluorite superstructure that is missing one fourth of the anions, allowing for small rearrangements of the ions. A total of 16 of these "structural oxygen vacancies" are situated at the *c* sites, according to international notation [15]. Depending on the location of these structurally vacant anion sites, two non-equivalent cation sites exist. Figure 1 shows a schematic representation of the two distinct cation sites, which are referred to as equipoints *b* and *d*.

Eight *b* site cations have six equidistant oxygen anion neighbors at 2.18(1) Å. These oxygen anions lie approximately at the corners of a distorted cube with two anions vacancies along one body diagonal. The remaining 24 cations are found at the more distorted *d* sites, where the two anion vacancies lie along a face diagonal of the distorted cube. There are three cation$_d$ to oxygen bond lengths: 2.13(1) Å, 2.19(1) Å and 2.23(1) Å, which average 2.18(1) Å. The remaining 48 atoms are lattice oxygen anions located at the general position (*e* site) and coordinated to four indium cations.

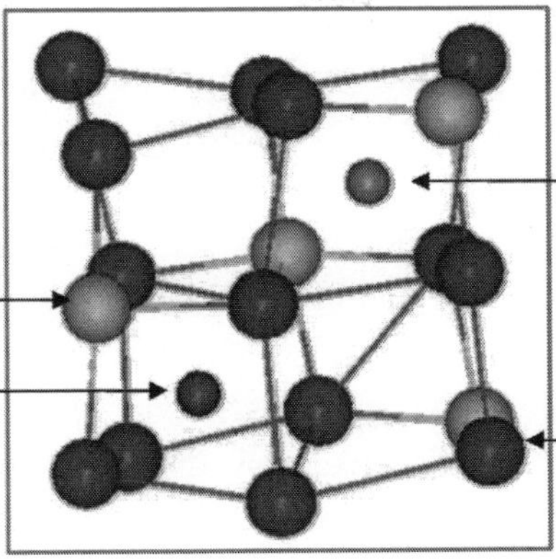

Figure 1. Non-equivalent *b* and *d* cations sites found in bixbyite In_2O_3.

The defect structure of pure In_2O_3 was indirectly deduced from electrical measurements performed by De Wit *et al.* [16,17,18,19]. They observed a −1/6 dependence of log (n) (carrier concentration) as a function of log (pO_2) (partial pressure of oxygen). The non-stoichiometric decomposition and its corresponding equilibrium constant are described below:

$$2In_{In}^{x} + 3O_{O}^{x} \Leftrightarrow 2In_{In}^{x} + 3V_{O}^{\bullet\bullet} + 6e' + \frac{3}{2}O_2 \qquad (2)$$

$$K_{vac} = [V_{O}^{\bullet\bullet}]^3 n^6 P_{O_2}{}^{3/2} \qquad (3)$$

The electroneutrality condition is

$$n = 2[V_{O}^{\bullet\bullet}] \qquad (4)$$

Substituting Equation (4) into Equation (3), the carrier concentration dependence on pO_2 is

$$n \propto P_{O_2}{}^{-1/6} \qquad (5)$$

Therefore, the observed −1/6 slope indicated that undoped In_2O_3 exhibits n-type conductivity due to the presence of oxygen vacancies under reducing conditions. These vacancies would be formed by removing some of the lattice anions located at the *e* sites and therefore introducing non-stoichiometry in the sample to obtain In_2O_{3-x} where *x* depends on the oxidizing conditions of the material, but it is usually less than 0.01 [16]. Due to the small population of oxygen vacancies that is predicted in pure indium oxide (less than 0.33% of the lattice oxygen site), no x-ray or neutron diffraction experiment has been able to confirm their existence. XRF techniques cannot determine the In/O ratio with such high precision either. However, theoretical models agree that the prevailing defects causing n-type conductivity in pure indium oxide are oxygen vacancies [5].

3.2. Bixbyite ITO

Indium-tin oxide is obtained by doping In_2O_3 with tin. The bixbyite structure of the parent indium oxide is kept up to the maximum solubility limit, and for temperatures lower than 1345 °C in samples at thermodynamic equilibrium [20,21]. Frank and Köstlin [4] pioneered the study of defect models for ITO. They proposed different clusters of defects to fit observed electrical properties. A summary of their model is described in the following section.

3.2.1. Frank and Köstlin Model for ITO

Resistivity, carrier concentration, and mobility were measured at room temperature as a function of Sn-doping and pO_2 on ITO thin films. Under reducing conditions, Frank and Köstlin observed that for all doping levels, a −1/8 slope on log (n) *vs.* log (pO_2) was followed by a plateau. The transition at which the carrier concentration plateaued depended on the Sn content. The carrier content increased with tin doping level up to a certain Sn concentration (~5 cation%, depending on pO_2), and then decreased. On the increasing part of the curve, the slopes became increasingly shallow with increasing Sn doping, *i.e.*, the more Sn, the less efficiently the electron population increased.

At doping levels <1 cation% and very high reducing conditions (~10^{-20} bar), the observed carrier concentration was proportional to the tin doping level C_{Sn}. Such a behavior would indicate that every

Sn•donor contributes one electron when the material is highly reduced. However for tin doping levels higher than 1 cation%, regardless of pO_2, the carrier concentration to tin concentration dependence deviated to slopes smaller than 1. In order to fit their observations, they postulated the existence of various Sn-$O_i$clusters.

The −1/8 slope of log (*n*) *vs.* log (pO_2) was explained from the disassociation of reducible $(2Sn^{\bullet}O_i'')^x$clusters. According to this model, the two tin ions are not nearest neighbors, and when the partial pressure of oxygen decreases it is possible to ionize this cluster, as shown in Equation (6).

$$(2Sn^{\bullet}O_i'')^x \leftrightarrow 2Sn^{\bullet} + 2e' + \frac{1}{2}O_2(g) \tag{6}$$

$$K = \frac{[Sn^{\bullet}]^2 n^2 {P_{O_2}}^{1/2}}{[(2Sn^{\bullet}O_i'')^x]} \tag{7}$$

Assuming that oxygen vacancies are minority species, the electron concentration is equal to the concentration of Sn•. Therefore, the carrier concentration dependence agrees with experimental observations. When the Sn doping level is small, n∝CSn14/ and the total Sn concentration in the sample can be written as

$$C_{sn} = [Sn^{\bullet}] + 2[(2Sn^{\bullet}O_i'')^x] \tag{8}$$

The n∝CSn14/ dependence agrees with the experimental observations at oxidizing conditions (<10^{-2}bar) and at low tin doping levels (<4 cation%). However, the presence of both reducible clusters and Sn• species explained only some of the observations. For very highly reducing conditions and doping levels greater than 1 cation%, the observed carrier concentration vs. doping concentration exhibited slopes smaller than 1, and the slopes decreased with increasing pO_2. Frank and Köstlin proposed the existence of an "irreducible" (2Sn·Oi''3OO)x cluster. The cluster also has 2 non-nearest tin neighbors that tightly bind to one oxygen interstitial and to three nearest lattice oxygens.

At high Sn concentrations (>5 cation%), adding more tin decreased, rather than increased, the electron population. Therefore, Frank and Köstlin hypothesized the presence of another cluster formed by the aggregation of a reducible and an irreducible cluster.

$$(2Sn^{\bullet}O_i'')^x + (2Sn^{\bullet}O_i''3O_O)^x \leftrightarrow ((2Sn^{\bullet}O_i'')^x(2Sn^{\bullet}O_i''3O_O)^x) \quad (9)$$

The total Sn concentration is then expressed as

$$C_{Sn} = [Sn^{\bullet}] + 2[(2Sn^{\bullet}O_i'')^x] + 2[(2Sn^{\bullet}O_i''3O_O)^x] + 4[((2Sn^{\bullet}O_i'')^x(2Sn^{\bullet}O_i''3O_O)^x)] \quad (10)$$

Equation (10) has a mixture of effective and non-effective tins. The non-effective species act as a sink of electrons, while the effective tins actually donate electrons. The total effective Sn concentration is defined as:

$$C_{Sn_e} = [Sn^{\bullet}] + 2[(2Sn^{\bullet}O_i'')^x] + 2[((2Sn^{\bullet}O_i'')^x(2Sn^{\bullet}O_i''3O_O)^x)] \quad (11)$$

3.2.2. Defect Structure Study of ITO Using X-ray and Neutron Techniques

A comprehensive study on the defect structure of indium-tin oxide (ITO) was performed by the author [22,23,24]. Powder samples of three tin concentrations (0%, 3%, and 9% Sn) were prepared under oxidizing and reducing conditions. Electrical conductivity and thermopower were measured and the samples were also characterized using EXAFS, anomalous x-ray diffraction, neutron diffraction, and XRF techniques.

Undoped In_2O_3 and 3% doped ITO bulk samples were sintered at 1350 °C in air. The samples were subsequently annealed in air at 800 °C for 15 h and quenched to room temperature. An extra annealing step was performed for some samples, either at 800 °C in CO/CO_2 ($pO_2 \sim 10^{-14}$ atm) for 65 h or at 500 °C in forming gas (4% H_2/96% N_2) for 6 h. Nano-ITO powders with 9% Sn were sintered at 700 °C for 1 h. The nano samples were then annealed in air for 5 h at 500 °C and quenched to room temperature. Some of these samples were also reduced in forming gas at 500 °C for 6 h and quenched to room temperature.

In situ conductivity and thermopower measurements were performed at 800 °C for the ITO bulk samples and at 500 °C for the nano-ITO samples. In agreement with Frank and Köstlin [4], a −1/8 slope of log (conductivity) *vs.* log (pO_2) was followed by a plateau at highly reducing conditions. Furthermore, a 1/24 slope of normalized thermopower *vs.* log (pO_2) was also followed by a plateau. The observed pO_2dependences of conductivity and thermopower support the ionization of reducible clusters, while the plateau regime is consistent with the existence of irreducible clusters.

EXAFS measurements were performed at the 5-BMD beamlime of the Advanced Photon Source (APS). The experiments were performed in transmission mode at the In *K* absorption edge (27,940 eV) and at the Sn *K*absorption edge (29,200 eV). The measured signal ranges for the In edge were from k = 3.3 to 14 Å^{-1} and from 3.3 to 12.6 Å^{-1} at the Sn edge. The indium to oxygen distances in bulk and nano-ITO samples were 2.18 to 2.19 Å, which are similar to the In-O distance in undoped In_2O_3 (2.18 Å). The first coordination shell consisted of approximately 6 neighbors in all samples. The tin to oxygen distances ranged from 2.05 to 2.08(1) Å in ITO samples, which was shorter by approximately 0.12 Å compared to In-O. This result agreed with EXAFS experiments performed by other research groups [25,26,27] and was expected based on the 0.10 Å radial difference between Sn^{4+} and In^{3+} ions. As a comparison, in SnO_2, the Sn-O distance is 2.045 Å. The first Sn-O coordination shell in ITO consisted of approximately 6 neighbors; however, due to the 10% uncertainty in this number, it was difficult to accurately distinguish a coordination number of 6 from 7. However, the EXAFS result definitely suggested that oxygen anions moved closer to tin than to indium in all of the ITO samples.

Anomalous x-ray diffraction experiments were performed at the APS (5-BMC beamline). High-resolution, x-ray diffraction patterns on all quenched samples were measured at room temperature below and at the In*K* absorption edge to maximize the contrast between indium and tin. Time-of-flight neutron powder diffraction patterns were also measured at room temperature, using the Special Environment

Powder Diffractometer, located at (alas, shuttered) the Intense Pulsed Neutron Source at ANL. The measured *d* ranges for the x-ray data were 0.87 to 4.13 Å and 0.50 to 3.00 Å for the neutron data. Both x-ray and neutron data sets were refined simultaneously using the Rietveld method [28].

The overall tin concentration in the samples was determined by XRF and confirmed by Rietveld analysis. The nano-powders were phase-pure, while the bulk samples had 96.5 wt % ITO and 3.5 wt % $In_4Sn_3O_{12}$. The nano-powders had approximately 9% Sn, which was beyond the solubility limit. This extended solubility was attributed to the metastability of nano-particles, where more tin can be incorporated in the structure. The combined analysis of diffraction patterns showed that in ITO tin substitutes into both *b* and *d* cation sites but with a preference of the *b* cation site, which has higher symmetry (see Figure 1). This result was in agreement with Mössbauer experiments [27]. The percentage of tin occupation in the *b* sites for bulk-ITO ranged from 5(3) to 9(3)%, while for the *d* site it was 1(1) to 2(1)%. For the nano-ITO samples, the tin occupancies were 15(4) to 25(5)% and 3(1) to 7(1)% for the *b* and *d* sites, respectively.

All ITO samples exhibited the presence of oxygen interstitial atoms located at the *c* sites, previously referred to as the anion structural vacancy site in bixbyite. On the other hand, for the undoped In_2O_3 samples the *c* site remained vacant. The amount of O_i present varied from sample to sample, but consistently the oxidized samples had higher O_i populations than the reduced samples. For bulk-ITO, the O_i occupational percentages in the samples were: 2.6(5)% when oxidized; 2.1(5)% after forming gas reduction; and 1.0(5)% after CO/CO_2 reduction. For the nano-ITO samples, the O_i populations were higher: 10.4(8)% and 5.7(7)% under oxidizing and reducing conditions, respectively. In a 2 × 2 × 2 fluorite superstructure the equivalent position of the *c* site would be $x = y = z = 0.125$ fractional lattice units. The refined positions of the O_i site ranged from 0.80(5) to 0.086(3) for bulk-ITO and from 0.085(2) to 0.091(1) for nano-ITO. The O_i position in reduced samples was consistently smaller for both bulk- and nano-ITO compared to the oxidized specimens.

The ratio of total Sn to O_i was calculated for all samples and compared to the value of 2 proposed by Frank and Köstlin for both reducible and irreducible clusters. For bulk oxidized ITO, the Sn/O_i ratio was 2.2(4); and upon reduction in forming gas, it increased to 2.7(6). Because the CO/CO_2 treatment dramatically decreased the O_i population, the ratio for this sample was 5.5(2.4). For nano ITO, the Sn/O_i ratio was 1.7(1) and upon reduction it became 3.1(3). The close proximity of the experimental values in oxidized ITO to the ratio of 2 supported the idea that the substitution of two $Sn^{\bullet}$ positively charged defects in the In_2O_3 lattice bring in an additional O_i'' species to charge compensate, as in the reducible $(2Sn^{\bullet}O_i'')^x$ cluster. If all the Sn and O_i species were in reducible clusters, upon reduction all those complexes would be ionized resulting in an increase in conductivity and in the total removal of O_i defects. However, since the highly reduced sample contained some O_i population and a plateau in conductivity, the presence of irreducible clusters was also supported.

Theoretical models [3,29,30] propose that the non-reducibility in ITO is due to the aggregation and proximity of tin cations around oxygen interstitials. As mentioned earlier, the refined O_i positions (0.80 to 0.91) were smaller than in the fluorite structure (0.125). In bixbyite, the ideal position would correspond to 0.116 atomic lattice coordinates, such that an oxygen interstitial is located 2.35 Å from 1*b* and 3*d* cations, as shown in Figure 2. The smaller experimental coordinates corresponded to a displacement of the O_i species towards the *d* site plane and away from the *b* cation. The distance from O_i to *d* sites was 2.2(1) Å, while the distance to the *b* cation increased to 2.8(1) Å. Theoretical calculations showed that isolated O_i^x defects would be located at 0.110 fractional coordinates, which is far from the experimental value. However, when clustering of tin and oxygen interstitials was modeled, the agreement with experiments dramatically improved. These theoretical models also predicted that tin strongly preferred the *d* site in clusters where only nearest neighbor O_i species participate. When looking at second and higher coordination shells, then tin preferentially substituted into the *b* site, in agreement with the experimental observations.

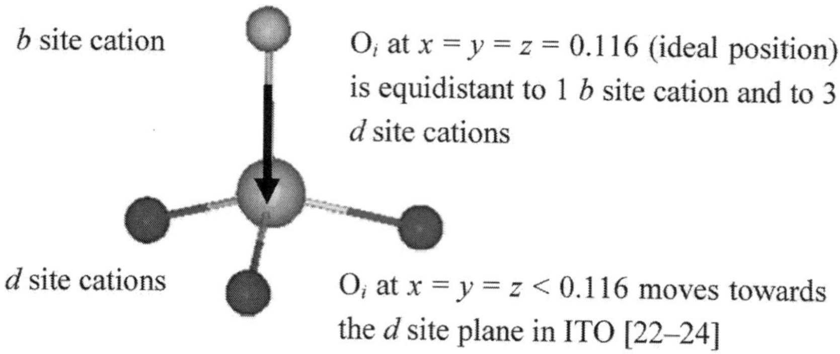

Figure 2. Nearest cations surrounding an oxygen interstitial located at the *c* sites in the bixbyite structure.

Other theoretical studies have tried to explain the *b* site preference that was experimentally observed by proposing that the synthesis conditions may shift the tin distribution in the indium sites [31]. Further theoretical investigations have corroborated that the nonreducibility of defects was due to the aggregation of tin cations around oxygen interstitials, but that also the chemical potential of conduction electrons increased with decreasing O_i population, thus limiting the maximum concentration of electrons available for conduction [32].

Rietveld refinements also revealed that the bixbyite lattice parameter expanded with increasing tin concentration. Considering that the radius of Sn^{4+} is 0.1 Å smaller than In^{3+}, cell contraction was expected with tin substitution. Similar increases in the ITO lattice parameter have been reported in the literature [4,21,27], and attributed to an incomplete charge compensation of $Sn^{\bullet}$ defects by O_i'' defects. As more positively charged tin defects are created, the repulsion among these species increases resulting in cell expansion. This explanation was consistent with the further expansion that was observed in reduced samples, where less O_i'' species were present enhancing the repulsion.

3.3. Zinc and Tin Doped Indium Oxide (ZITO)

Another indium-based transparent conducting oxide results from co-doping indium oxide with zinc and tin. The material is referred to as ZITO and is a relatively new TCO material. It is an attractive substitute to ITO since zinc and tin replace indium, which has become more expensive and scarce. Since the first literature report of ZITO in 1995 [33] many studies on the properties, range of compositions, and defects have been published [34,35,36,37,38,39,40,41,42,43,44,45,46,47,48,49]. N-type bixbyite ZITO materials having more tin than zinc exhibit electrical properties comparable to ITO. Recently, ZITO's defect structure has been studied using crystalline powders [43,44], amorphous powders [46], and amorphous thin films [47,48]. Crystalline ZITO can exhibit the bixbyite or the corundum phase. For the corundum phase, up to 70% co-substitution is possible [44], while according to Palmer *et al.* [35] up to 40% of the indium can be replaced with Sn^{4+} and Zn^{2+} in bixbyite ZITO. The preparation of the corundum phase requires a more complicated process since high pressure (~70 kbar) and high temperature (1000 °C) are required to transform the material [44].

3.3.1. EXAFS Studies of Crystalline ZITO (c-ZITO) with the Bixbyite Structure

3.3.1.1. Study on Bixbyite c-ZITO Powder Materials

The local structure of bixbyite ZITO was investigated by Hoel *et al.* [43] on powders using EXAFS. The materials studied were $In_{2-2x}Zn_xSn_xO_3$ (x = 0.1, 0.2, 0.3, and 0.4) and were referred to as ZITO-10, ZITO-20, ZITO-30, and ZITO-40. The measurements were performed at the APS in transmission mode at the In, Sn, and Zn *K* absorption edges. Because EXAFS experiments do not have enough resolution to distinguish the *b* and *d* cation site environments, the results provided an average of distances for each coordination shell. Therefore, it was not evident whether the dopants substituted for In randomly or if there was a site preference as determined in the ITO defect structure study [22,23,24]. Hoel *et al.* [43] assumed that tin and zinc substituted randomly in the indium sites. Laboratory XRD measurements confirmed the phase purity of the samples and indicated that the In_2O_3 bixbyite unit cell contracted linearly as tin and zinc co-substitutions increased, consistent with the smaller radii of the dopants.

Similar EXAFS features were observed for ZITO and In_2O_3 bixbyite powders at the In edge. The main difference was the decrease of the amplitudes for both the In-O and In-In/Sn/Zn bonds in ZITO as the parameter *x* increased. This decrease was attributed to the weaker scattering of Zn which has 30 electrons compared to In and Sn, which have 49 and 50 electrons, respectively. Furthermore, Zn interfered destructively with the In and Sn EXAFS signals due to a phase shift resulting in an overall weaker signal. The coordination number for all 4 measured In shells was fixed to 6. The In-O distance for the first shell averaged 2.161(4) Å for all ZITO compositions, compared to 2.18 Å in In_2O_3. The next three shells corresponded to In-In/Sn/Zn. For these cation shells higher co-substitution consistently resulted in shorter bond lengths. For the second shell, the distances ranged from 3.348(5) Å for ZITO-10 to 3.321(7) Å for ZITO-40. For the third shell and fourth shell, the bond lengths ranged from 3.820(5) Å to 3.789(8) Å, and from 5.068(7) Å to 5.027(10) Å, respectively. These observations were consistent with the unit cell contraction measured from XRD and with the smaller radii of Zn^{2+} and Sn^{4+} compared to In^{3+}.

EXAFS data for ZITO-20, ZITO-30 and ZITO-40 were reported at the Sn edge. EXAFS results indicated that the tin local environment for bixbyite ZITO was similar to that of In in the bixbyite structure. The same structural features were observed, however, the amplitude of the ZITO spectra again decreased with increasing *x*, due to destructive interference from Zn. The first four coordination shells were fixed to 6 neighbors. The Sn-O first shell distance contracted owing to the smaller Sn^{4+} radius, and it was actually closer to the bond in SnO_2 (2.05 Å) than to the bond in In_2O_3 (2.18 Å) or in ZITO at the In edge (2.16 Å). The Sn-cation shells were also shorter than the In-cation shells. However, unlike the In shells, they showed a very modest bond-length shortening with increasing co-substitution, and the actual distance ranges were almost independent of composition within experimental uncertainties. The range of distances were

2.080(7) Å for ZITO-20 to 2.073(6) Å for ZITO-40 for the first shell, 3.36(1) Å to 3.34(1) Å for the second shell, 3.83(1) Å to 3.81(1) Å for the third shell, and 5.02(1) Å to 4.99(1) Å for the fourth shell.

The coordination type in zinc was determined using XANES measurements at the Zn edge. The Zn-O shell in ZnO is tetrahedral, and as described before, in bixbyite the In-O shell is octahedral. XANES experimental results indicated that the local environment was octahedral as in bixybite In_2O_3, so the first three zinc coordination shells were fixed to 6 neighbors. As in the Sn edge, data for ZITO-20, ZITO-30 and ZITO-40 were reported. The Zn-O first shell distance also contracted owing to the smaller Zn^{2+} radius, and within experimental uncertainty, it was the same as the Sn-O shell. The range of distances were 2.084(7) Å for ZITO-20 to 2.075(6) Å for ZITO-40 for this first shell. Unlike the In results, higher coordination shells did not seem to shorten with increasing co-doping taking into account the larger experimental uncertainties at the Zn edge. The average distances were 3.30(1) Å and 3.74(2) Å for the second shell and third shells, respectively. The higher uncertainties were attributed to a more disordered local environment.

3.3.1.2. Study on c-ZITO Thin Films

Proffit *et al.* [47,48] studied the local structure of a bixbyite ZITO thin film deposited on a sapphire substrate. The sample had an In-Sn-Zn composition of 78-12-10 as determined using XRF in a scanning electron microscope. The conductivity was 1565 S/cm; the Hall mobility was 41 cm^2/Vs; and the carrier concentration was 2.4×10^{20} electrons/cm^3. EXAFS and XANES measurements were performed at the In, Zn and Sn *K*-edges using a Ge solid-state detector in fluorescence mode using three different orientations. The incident angles were 6°, 30° and 50° with respect to the sample surface. The data were collected at the 5-BMD beamline of the APS. The EXAFS data for k = 3.2 to 10 $Å^{-1}$ and R = 1.0 to 4.1 Å were fit.

XANES spectra indicated that the oxidation states of the cations were Zn^{2+}, In^{3+}, and Sn^{4+}. Furthermore, all the cations exhibited octahedral coordination, similar to the results in crystalline bixbyite ZITO [43].

The In-O bond length was 2.17(3) Å with 5 to 6 neighbors depending on the incident angle. The Sn-O bond was split 2.08/2.25(3) Å with a total of approximately 6 neighbors. The Zn-O bond was also split 2.04/2.18(3) Å with a total of approximately 5 neighbors.

The cation-oxygen average bond lengths were similar to the In-O bond in bixybite (2.18 Å). The uncertainty in the coordination numbers was conservatively on the order of 10%, so taking into account these uncertainties it was difficult to distinguish 5 from 6 neighbors. However, because for all 3 orientations of the films the results consistently showed that the zinc cation was under-coordinated, the authors postulated a redistribution of oxygen near Zn. One lattice anion displaced to a *c* site near In and Sn can leave behind an oxygen vacancy. The presence of an additional oxygen vacancy around Zn would result in 5 neighbors instead of 6. While the presence of an extra oxygen around In and Sn would increase the coordination number of these cations to 7, the overall stoichiometry and charge balance would remain the same. However the local environment around each of the cations would be different. Measurements on ZITO samples by Harvey *et al.* [42] found the same −1/8 slope of log (conductivity) *vs.* log (pO_2) that was reported before for ITO materials [4,22,23,24]. The defect mechanism responsible for the n-type behavior in ZITO therefore has been attributed to Frank and Köstin defects similar to those in ITO [40]. Samples with excess Sn (~2.5%) result in increased conductivities compared to samples with excess Zn (~2.5%) or with nominal [Sn] = [Zn] compositions. The results therefore indicated that ZITO samples with excess aliovalent Sn (Sn > Zn) exhibited donor populations when reduced, as in the case of ITO. Assuming that the Zn dopants were completely compensated by isovalent Sn species, Proffit [48] calculated that the remaining 2% of Sn would result in conductivities comparable to those measured. The zinc cations therefore would not contribute to the conductivity but would allow for higher levels of indium substitution in ZITO compared to ITO.

3.3.2. Defect Structure Study of Amorphous ZITO (a-ZITO) Using X-ray Techniques

Amorphous ZITO samples are attractive due to their high electrical mobilities compared to crystalline materials. The lack of grain boundaries reduces the scattering of charge carriers. When amorphous ZITO samples are heated, the electrical conductivity increases and reaches a maximum just before crystallization occurs. The carrier concentration seems to remain unaffected, so the increase in conductivity has been attributed to the mobility [49]. In amorphous samples, the definition and the role of defects is less apparent than in crystalline samples. It is clear that unlike crystalline materials, amorphous samples lack long-range order; however, they seem to exhibit short-range order as reported by studies on amorphous ZITO powders and thin films, which are described next.

3.3.2.1. Study on a-ZITO Powder Materials

A comprehensive investigation of the defect structure in a-ZITO powders was performed by Hoel *et al.*[46], and a summary of their results is described here. The samples were $In_{2-2x}Zn_xSn_xO_3$ where x = 0.2, 0.3, and 0.4. The powders were synthesized using a precipitation and calcination process. The characterization techniques included XRD, EXAFS, XANES, total scattering PDF, and transmission electron microscopy (TEM). The authors compared the results of the a-ZITO with those of crystalline bixbyite and corundum In_2O_3. TEM images revealed that 20 to 50 nm particles aggregated in a random manner. Diffraction data were consistent with an amorphous arrangement of atoms.

High resolution XRD experiments were performed at room temperature at the 11-BMC beamline of the APS. The diffraction pattern consisted of broad peaks whose positions corresponded to approximately the *d*-spacings of crystalline bixbyite and corundum. *In situ*, high temperature, XRD measurements, were performed at the 5-BMD beamline of the APS using a monochromatized beam of 0.654 Å. Data on the precursors were collected from room temperature up to 568 °C in transmission mode using a 2D detector to study the crystallization process. At room temperature only the crystalline precursor diffraction peaks were present. At 300 °C the signal became completely amorphous. As the sample was heated to even higher temperatures, the diffraction peaks gradually sharpened. After heating to ~560 °C, the sample crystallized into a mixture of bixbyite and corundum phases.

EXAFS data were measured at the In, Sn, and Zn *K* edges in transmission mode at the APS. The amount of zinc and tin did not have a significant effect on the EXAFS results at the In, Sn and Zn edges. Several constraints were applied to the data at the In edge: only In neighbors were considered in the fits; the In-O shell coordination number was set to 6 as in bixyite and corundum In_2O_3; the coordination number for edge- and corner-sharing cations were set to be an average of 6 for each case. The In-O distance was 2.140(4) Å, slightly shorter than in bixbyite ZITO (2.16 Å). The second coordination shell was split and consisted of 3.2(4) edge-sharing indium neighbors at 3.28(1) Å, and 2.8(4) neighbors at 3.41(2) Å. The third shell was also split and consisted of 3.2(4) corner-sharing indium neighbors at 3.64(1) Å and 2.8(4) neighbors at 3.88(2) Å. The tin and indium local environments exhibited structural differences. Constraints were also applied at the Sn edge: only indium neighbors were fit for distances ranging from 3 to 4 Å; and the Sn-O coordination number was set to 6 as in rutile SnO_2. The Sn-O distance was 2.058(3) Å. The second, third and fourth shells consisted of Sn-In neighbors and were as follows: 4.1(4) neighbors at a distance of 3.27 Å; 3.0(2) cations located 3.48(1) Å apart; and 3.9(4) neighbors at 3.72(1) Å. For the Zn edge, XANES spectra were analyzed to determine if the coordination was either octahedral as in bixbyite and corundum ZITO or tetrahedral as in ZnO. The results indicated that for a-ZITO the Zn-O were tetrahedrally coordinated. EXAFS experiments showed that the Zn-O shell was 1.97(2) Å when the number of neighbors was fixed to 4 in a-ZITO. This distance agreed with tetrahedral coordination. No shells beyond 2 Å were observed; this behavior differed from that of the crystalline bixbyite and corundum phases.

PDFs measurements were collected in transmission mode at 6-ID-D of the APS at ANL using 87.06 keV x-rays and a 2D detector. Relatively sharp peaks corresponding to atomic pairs of less than 5 Å were obtained for all compositions. The amount of zinc and tin did not seem to have a significant effect on the PDF results, consistent with the EXAFS observations. The first PDF peak at 2.14 Å corresponded to an average of In-O, Sn-O, and Zn-O pairs. The second and third peaks had similar integrated intensities and occurred at 3.29 Å and 3.73 Å corresponding to edge and corner-sharing cation-cation pairs, respectively. At distances higher than 5 Å, the peaks for a-ZITO became broader indicating a decrease in order. For comparison purposes, the authors measured PDFs of bixbyite and corundum In_2O_3 and found that those crystalline samples exhibited sharp peaks up to 20 Å (the maximum range of measured results).

From the above experiments the authors concluded that a-ZITO had structural differences with bixbyite and corundum ZITO. In the crystalline phases, all the cations are octahedrally coordinated to oxygen; however the Zn environment exhibited tetrahedral coordination in the amorphous phase, which probably prevented crystallization of the samples. The In-In and Sn-In local environments differed from each other, while the Zn-In environment seemed to be too disordered to exhibit any measurable structure. At 568 °C, a-ZITO crystallized into a mixture of corundum and bixbyite phases. At 700 °C, both crystalline phases still coexisted; however, higher zinc and tin substitutions favored the presence of corundum.

3.3.2.2. Study on a-ZITO Thin Films

Proffit [48] studied the local structure of several amorphous ZITO thin films. A set of samples with a Sn/Zn ratio of 12/14 (actual ratio 11.9/13.5, also referred to as ZITO30, since approximately 30% of indium is replaced with Sn and Zn) was prepared to study the effect of pO_2 during growth. The pO_2 values were 0.5 mTorr, 7.5 mTorr, and 15 mTorr. The corresponding conductivities were 7, 1457, and 342 S/cm; the Hall mobility values were 1.4, 49.5 and 53.0 cm^2/Vs, while the electron concentrations were 2.9×10^{20}, 1.7×10^{20}, and 3.7×10^{19} cm^{-3}. Since the 7.5 mTorr conditions resulted in the best electrical properties, two other compositions were prepared at this pO_2: 16/14 (actual ratio 15.5/13.6) and 18/13 (actual ratio

17.8/13.3). The corresponding conductivities for these samples were 1226 and 1431 S/cm, while the Hall mobilities were 48.4 and 46.7 cm^2/Vs, and the electron concentrations were 1.6 × 10^{20} and 1.9 × 10^{20} cm^{-3}. The electrical properties seemed to be optimized for medium pO_2 environments (7.5 mTorr). However, the Sn/Zn ratio did not have an effect on the electrical performance of the amorphous samples as it did in the bulk crystalline samples, where higher In compositions resulted in increased conductivities for Sn-rich samples (~2.5%) [40]. It should also be noted that the 12/14 Sn/Zn composition film had an excess of zinc, but it resulted in n-type conductivity with high carrier concentration and mobility, unlike bulk crystalline samples, where Zn-rich samples exhibited n-type behavior with inferior electrical properties. Another sample having a total of 70% of indium replaced with zinc and tin was prepared at 7.5 mTorr. This composition in crystalline samples was well beyond the solubility limit, which is about 40%.

As in the crystalline thin film sample of composition 12/14, XANES measurements for most samples, indicated that the cations present in a-ZITO were Zn^{2+}, In^{3+} and Sn^{4+}. The only sample where Sn^{2+} might have been present was the sample prepared at 0.5 mTorr. The In and Sn XANES signals were very similar for the amorphous and crystalline samples. However, the Zn signal was broader due to a higher degree of disorder. The radial distribution functions showed that only first shell ordering can be measured in amorphous samples. The In-O bond length was 2.14(2) Å with a coordination number of 5.0(4). The bond length was similar to those of the crystalline film and powders within experimental uncertainty. The Sn-O bond length was 2.07(3) Å, which matched the shorter distance in the crystalline film. This distance was also consistent with the amorphous powders and with the length in SnO_2. The coordination number in a-ZITO films was 5.9(3), except for the sample prepared at 0.5 mTorr, whose coordination number was 4.1(3). The Zn-O bond length was even shorter, 1.98(1) Å, also comparable to the a-ZITO powders and to the length in ZnO. The number of oxygen neighbors in the a-ZITO films was 3.3(2), which is much smaller than in the c-ZITO thin film.

Anomalous PDF measurements on a modest data range up to q = 9 $Å^{-1}$ were performed at the Zn *K*-edge on a film similar to the 12/14 composition. While the uncertainties of these results were large due to the limited measured range, the PDF observations were consistent with the EXAFS results for the Zn local environment. Similar trends in the coordination numbers for the first shell in c-ZITO and a-ZITO films were observed, with Zn having fewer neighbors than In and Sn. The role of Zn in a-ZITO was less evident compared to c-ZITO where it served to compensate some of the Sn species and allowed for higher doping levels. In amorphous samples, the definition and the role of defects is less apparent than in crystalline samples. In the case of ZITO the Sn/In ratio had no effect on the electrical properties, while in crystalline samples the composition had a major effect attributed to different populations of defects. ZITO samples crystallized at higher temperatures (300–345 °C [48]) compared to ITO (185–230 °C [50]) and In_2O_3 (165–210 °C [50,51]), therefore zinc may play a key role in increasing the stable temperature range for a-ZITO.

As mentioned before, crystallization of amorphous samples typically causes a reduction in the mobility of the samples, due to the increased scattering of carriers by grain boundaries. Buchholz *et al.* [49] reported that the conductivity of amorphous samples increased with temperature and then decreased when crystallization occurs. The authors attributed the enhanced conductivity during annealing to the homogenization of defect populations and to the strain relaxation in the sample. Proffit [48] studied the crystallization of ZITO films using *in situ* glancing incidence x-ray diffraction (GIXRD). Consistently, the addition of zinc to the sample resulted in higher crystallization temperatures. Samples with composition 12/14 (ZITO30) that were deposited at room temperature and 5 mTorr O_2 crystallized between 300 and 345 °C. In another experiment, the effect of substrate temperature during deposition at 5 mTorr was also explored on ZITO samples with 30%, 50% and 70% co-doping levels. The temperatures measured were 25 °C, 200 °C and 400 °C. For ZITO30 crystallization started at a substrate temperature of 200 °C, while for ZITO50, the temperature was 400 °C. The ZITO70 sample remained amorphous at 400 °C, however the substrate temperature was not increased further to prevent the evaporation of zinc. The optical electrical properties of ZITO50 were studied from −100 °C up to 400 °C, and it was determined that the optimal temperature for optical performance was 25 °C. However for this sample and at the experimental temperatures studied, at 200 °C the amorphous sample exhibited the best electrical properties. For ZITO30, the best electrical properties were at 25 °C, when again the sample was still in the amorphous phase. The ZITO30 sample was more conductive than the ZITO50 sample due to smaller carrier concentrations. These results suggested that composition, substrate temperature and further annealing modify the electrical and optical properties of a-ZITO materials, so it is possible to tune and optimize these properties to satisfy the demands for a range of applications.

3.4. Rutile Tin Oxide (SnO_2)

Tin oxide (TO) is an attractive transparent conductor since tin is less expensive and scarce compared to indium. Most of the applications of architectural glass which are deposited on building windows use tin-oxide based TCOs [52]. Rutile tin oxide crystallizes in a tetragonal structure with space group $P4_2/mnm$ (number 136) [53]. A unit cell with lattice constants a = 4.7374(1) Å and c = 3.1864(1) Å contains two tin and four oxygen atoms. The tin cations are located at the a sites, and the oxygen anions are found at the f sites, according to Wyckoff notation [15]. Edge-sharing octahedrally coordinated tin ions form chains along the [001] direction. From the oxygen perspective, the tin atoms are located at approximately the corners of an equilateral triangle. Figure 3 shows a representation of a unit cell of rutile TO.

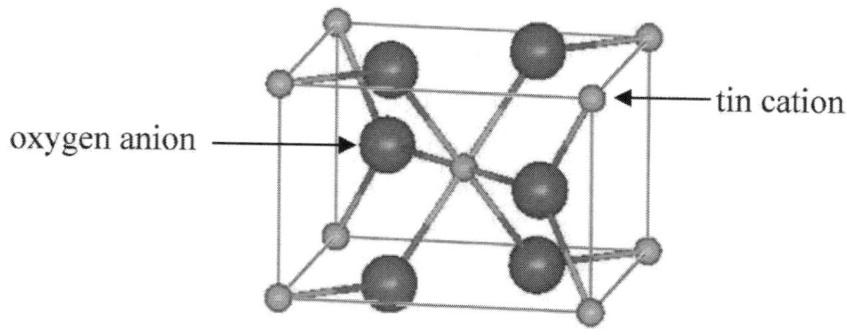

Figure 3. One unit cell of rutile SnO_2.

Samson and Fonstand [54] established that TO exhibits n-type conductivity. They observed approximately a −1/6 slope in a log (conductivity) as a function of log (pO_2), similar to the behavior observed for pure In_2O_3. Thus, they concluded that the major defect species in SnO_2 consists of oxygen vacancies (V··O). The non-stoichiometric decomposition and its corresponding equilibrium constant can be written as follows:

$$Sn_{Sn}^{x} + 2O_{O}^{x} \Leftrightarrow Sn_{Sn}^{x} + 2V_{O}^{\bullet\bullet} + 4e' + O_2 \quad (12)$$

$$K_{vac} = [V_{O}^{\bullet\bullet}]^2 n^4 P_{O_2} \quad (13)$$

The electroneutrality condition n=[2V··O] results in n∝PO2−16/. The material can then be expressed as SnO_{2-x} where x depends on the oxidation state. As in In_2O_3, the hypothesized population of oxygen vacancies responsible for n-type conductivity is relatively small, so it is not possible to detect it using XRD, neutron diffraction or XRF techniques.

EXAFS studies on undoped TO nano-powders [55,56,57] have studied the effect of particle size and disorder on short-range scales. Davis *et al.* [55] measured EXAFS and XRD on pure tin oxide nano-powders prepared by a sol-gel process. The particle size of 2 to 3 nm was determined from analyzing the peak width of the XRD peaks. The measurements at the Sn *K*-edge were conducted in transmission mode at the 9.3 station of the CLRC Daresbury Synchrotron Radiation Source. *In situ* experiments showed that the particles grew significantly at 400 °C. EXAFS spectra for higher coordination shells were reduced due to the small particle size. This observation was consistent with the results of Savin *et al.* [56] and Chadwick [57]. When the coordination numbers and bond lengths for nano-SnO_2 were simultaneously refined, both the distances and the number of neighbors were smaller compared to bulk-SnO_2, especially for higher coordination shells. For the first shell, the distance only shortened by 0.005 Å, and the coordination number was approximately 6. However, for second, third, and fourth shells, the coordination numbers decreased by a factor of 2, and the third Sn-O shell showed the most dramatic decrease (0.142 Å).

Savin *et al.* [56] prepared ball-milled and sol-gel nano-samples. The particle size was determined from analysis of the XRD peaks using a conventional Cu K_α source. EXAFS measurements were performed in transmission mode at station 9.3 of the CLRC Daresbury Synchrotron Radiation Source. The particle sizes obtained ranged from 8 nm up to 86 nm (this biggest size was referred to as "bulk" sample). Compared to the bulk sample, the EXAFS signals for the smaller nano-grains were attenuated, especially for the Sn-Sn bonds near 3 and 4 Å. The attenuation was inversely related to the particle size, except for the ball-milled sample with 16 nm, which showed the weakest signal of all samples. The EXAFS data were fit using two approaches. The first one refined the radial distances and Debye-Waller factors while fixing the coordination numbers to their crystallographic values: 6, 2, 4, and 8 for the first (Sn-O), second (Sn-Sn), third (Sn-O) and fourth (Sn-Sn) coordination shells, respectively. The radial distances refined to the ideal crystallographic values: 2.05 Å, 3.19 Å, 3.64 Å, and 3.70 Å, respectively, while the Debye-Waller factors increased with decreasing particle size and also with ball-milling time. The second refinement approach refined the coordination numbers and Debye-Waller factors while fixing the radial distances. In this case, the coordination numbers refined to smaller values due to a decrease in particle size. For the 11 nm sol-gel sample, the coordination numbers were 4.8, 1.5, 2.1, and 5.4, while for the 8 nm, sol gel sample they were 5.0, 2.4, 3.2, and 4.8. For the 16 nm ball milled sample, the coordination numbers were 5.0, 2.3, 4.4, and 3.6. For this last sample, the authors concluded that, besides grain size, the possible presence of an amorphous phase and an increase in disorder were important factors leading to the reduction in the number of neighbors. This conclusion was also derived from nuclear magnetic resonance results where an increase in structural disorder was evident.

Chadwick [57] also conducted EXAFS experiments at the 9.3 station of the CLRC Daresbury Synchrotron Radiation Source. Data were collected up to k = 18 Å^{-1} at the Sn K-edge on both bulk and 3 nm nanocrystals. Compared to the bulk sample, the EXAFS for the nanocrystalline sample was attenuated, and the signal from the Sn-Sn and bonds larger than 3 Å were dramatically reduced. This observation was consistent with the work of Savin *et al.* [56] and was attributed mostly to a reduced average coordination number and not disorder. Chadwick suggested that for grains smaller than 5 nm, due to a large fraction of surface atoms, the average coordination number of tin was reduced, although crystallites and interfaces could be highly disordered and contribute as well.

3.5. Sb-Doped SnO_2

Antimony-doped tin oxide (ATO) shows improved n-type conductivity compared to pure tin oxide. Berry and Laundry [58] performed Mössbauer experiments and found that at low doping levels antimony was primarily incorporated as Sn^{5+}. At increasing doping levels, more Sn^{3+} species were present but their incorporation decreased the conductivity [59]. Pyke *et al.* [60] used XRD, XRF, thermogravimetric techniques, and microscopy to study the solubility of antimony in TO at temperatures ranging from 600 °C to 1000 °C. The samples had starting compositions of $Sn_{1-x}Sb_xO_2$ for $0 \leq x \leq 1$. From the XRD peak widths, the average crystallite sizes increased from 4 nm to 18 nm. Equilibrium was difficult to achieve, and evaporation became problematic at higher temperatures and long annealing times. However, the results indicated that the solid solution of antimony in SnO_2 was less than 4 atomic percent. It should be noted that in thin films and nano-powders, higher Sb concentrations can be achieved. Several investigations on the defect structure of ATO consist of XRD [61,62], neutron diffraction [63], and EXAFS measurements [62,64,65]. A summary of those reports is presented next.

Gržeta *et al.* [61] used XRD and Mössbauer spectroscopy to investigate the substitution of Sb into the rutile SnO_2 structure using samples with different atomic doping levels: 0, 3.1(4), 6.2(6) and 11.9(5), and 14.0(7)% Sb. The concentrations in the samples were determined by atomic emission spectroscopy and particle induced x-ray emission. XRD measurements were performed using a laboratory Cu K_α source, and the data were analyzed using Rietveld refinement [28]. Analysis of the diffraction peak shapes indicated anisotropy in the nanosized grains. Mössbauer spectroscopy (^{119}Sn and ^{121}Sb)

results determined that tin was in the 4+ oxidation state, while antimony was in both the 3+ and 5+ states. The amount of Sb^{3+} was twice as much as the Sb^{5+} species. Structural results from XRD included atomic positions, isotropic Debye-Waller factors, fractional site occupancies, and lattice parameters. Both unit cell parameters, a and c, consistently increased with Sb-doping, except for the sample with highest doping level, where the lattice cell was only slightly larger than in pure tin oxide. All samples were phase pure, except for the sample with the most antimony, where 5 wt % of the sample was Sb_2O_3. For this sample the total amount of antimony was 19.4(7) atomic percent only 14.0(7) at.% was actually present in the rutile structure, and the rest was present in the secondary phase. The ratio c/a was 0.672 for all samples.

Compared to Sn^{4+}, Sb^{3+} is 0.07 Å larger and Sb^{5+} is 0.09 Å smaller in radius, so based on ionic size alone, an increase in the lattice parameters would indicate that Sb^{3+} substitutes for tin. Gržeta *et al.* [61] allowed the structural refinement of both 3+ and 5+ ions. Fractional occupancies were reported for the phase-pure samples. For the sample with 3.1%, the occupational fraction of both Sn^{4+} and Sb^{5+} added up to 0.96, while the occupation of Sb^{3+} was 0.04(2). For the sample with 6.2%, the fractional occupation of Sn^{4+} and Sb^{5+}species refined to 0.91, and to 0.09(2) for Sb^{3+}. The authors also refined the occupation of the oxygen site for these two samples. Because antimony has 51 electrons and tin has 50, in this case, the fractional occupancies obtained using a conventional Cu K_α source should be interpreted with caution, since, as mentioned in section 2.1, the oxygen signal for XRD is very weak, and elements that are next to each other in the periodic table are indistinguishable using a conventional Cu K_α. Anomalous synchrotron x-ray experiments and/or neutron diffraction measurements would provide more reliable results and insight into the substitution of antimony into the tin oxide rutile structure.

Berry and Greaves [63] used neutron diffraction to investigate the defect structure of 10.6(3)% ATO with 14 nm crystallites. The measurements were performed at room temperature, using the PANDA diffractometer at ERE, Harwell. The neutron wavelength was 1.5397 Å, and the scattering lengths for tin, antimony, and oxygen were 6.1 fm, 5.6 fm, and 5.8 fm, respectively. The data were analyzed using the Rietveld method [28]. Results included atomic positions, site occupancies, isotropic Debye-Waller factors, and lattice parameters. The cell parameters were $a = b = 4.7373(5)$ Å and $c = 3.1816(5)$ Å. The best refinement scenario fixed the oxygen occupancy to 2.0 atoms/cell and refined the tin occupation in the *a* site to 0.918(7). The model also included the presence of Sb in an interstitial position (*k* site) with an occupation of 0.11(1). The interstitial antimony was initially placed at $x = 0, y = z = 0.5$, the vacant octahedral sites of the rutile structure. The refined positions were actually shifted by 1.2 Å from these octahedral sites to $x = -0.055, y = 0.283, z = 0.326$ atomic coordinates. Figure 4 shows the location of those sites in rutile.

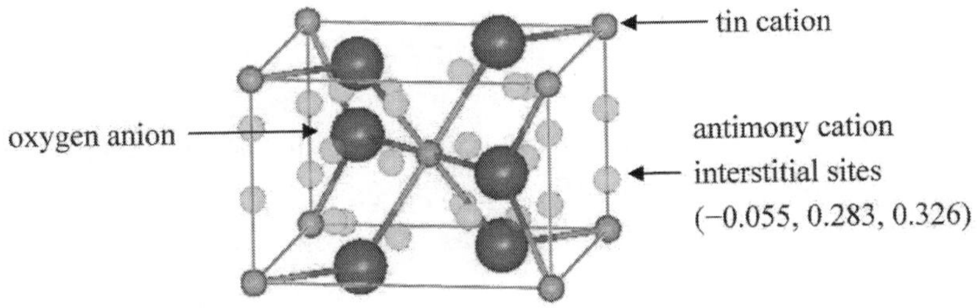

Figure 4. A unit cell of rutile SnO_2 showing the location of interstitial antimony refined by Berry and Greaves [63].

Berry and Greaves proposed a defect model that also includes oxygen vacancies to maintain charge balance and explain the n-type conductivity. For refinement purposes, the occupation of one species is always fixed and the rest of the sites are refined with respect to it. In the model, both cation and anion vacancies were proposed, so the relative occupancy ratios were maintained, but the absolute values were adjusted. The model then consisted of 1.90(1) O^{2-} anions, 0.871(7) Sn^{4+} cations, and 0.10(1) Sb^{3+} cations. The refined position of the antimony interstitial was in close proximity to anion vacancies. Such a Sb^{3+}-Vö pair would have a net charge of 5+, which could then balance the charge of negatively charged tin vacancies.

Rockenberger *et al.* [64] measured XANES and EXAFS on samples with 9.1 and 16.7 at% antimony, with particle sizes ranging from 2 to 6 nm. The samples were prepared by a co-precipitation method, and two different antimony chlorides were used $SbCl_5$ and $SbCl_3$ to study their effect on the antimony oxidation state distribution. XANES experiments were performed in transmission at the Sb L_1-edge at room temperature, while EXAFS measurements at the Sb K-edge were conducted at 5 K, at the E4 beamline of the DORIS synchrotron at HASYLAB, DESY. XANES spectra were compared to Sb_2O_3 (Sb^{+3}) and Sb_2O_5 (Sb^{+5}) reference materials to determine the oxidation state. The authors found that the ratio of Sb^{3+}/Sb^{5+} decreased with room temperature "aging", if the samples were dried only at 100 °C. These samples were non conductive and had a brownish, yellowish color. For the 9% Sb^{3+} sample, after 1 week at room temperature the Sb^{3+}/Sb^{5+} ratio was 35/65, and it decreased to 22/78 after sitting at room temperature for 6 months. For the 17% Sb^{3+} sample, the corresponding ratios were 50/50 and 21/79; while for the 17% Sb^{5+} sample, the ratio was 9.91 (and not reported after 6 months). If the samples were dried at 500 °C, then the ratios after 1 week and 6 weeks remained the same, their n-type conductivity increased and their color changed to blue. The ratios for the 9% Sb^{3+}, 17% Sb^{3+}, and 17% Sb^{5+} samples were: 28/72, 23/77 and 24/76, respectively, which were all very close to 1/3.

The EXAFS results by Rockenberger *et al.* [64] showed that the bond lengths did not change with sample composition or with drying processes. The bond lengths were: 2.00(2) Å, 3.20(2) Å and 3.71(2) Å, for the first shell (Sb-O), second shell (Sb-Sn/Sb), and third shell (Sb-Sn/Sb), respectively. These values were close to the 2.05 Å, 3.19 Å and 3.71 Å lengths in SnO_2. The coordination numbers changed slightly with sample composition and with anneals at 100 °C and 500 °C, however most of the changes were within one standard deviation. The first coordination numbers ranged from 4.1(3) to 5.4(4). The second shell coordination numbers ranged from 0.7(4) to 1.1(5). The third shell coordination numbers ranged from 1.3(6) to 2.4(10). The authors explained that the incorporation of antimony into the rutile tin oxide lattice of their co-precipitated samples using a two-step process. During the first step at low temperatures (100 °C), the stoichiometry of the non-conductive samples was given by SnxSbIIIySbVZO(2x+1.5y+2.5z). This composition required the metal: oxygen ratio to deviate from 1:2; and their results indeed ranged from 1:2.0 to 1:2.05. Upon annealing at 500 °C, the samples became conductive and their stoichiometry was either SnxSbIIIy−tSbVZ+tO2(x+y+z)·(z−y+2t)e−CB neglecting oxygen vacancies, or (Sn,Sb)O2−δ·(n+2δ)e−CB, where e−CB represented the number of conduction band electrons, and the number of oxygen vacancies per cell was given by δ. Since the samples were prepared in air, such oxidizing conditions would probably not produce a substantial population of oxygen deficiencies. However, the proposed models could not be confirmed due to the typical uncertainties associated with coordination number determinations using only EXAFS results.

Geraldo *et al.* [65] also investigated the incorporation of antimony in the rutile tin oxide lattice using EXAFS and XANES. The samples were nanocrystalline powders with compositions 0%, 1.5%, 3%, 4%, 9%, and 16% Sb obtained via a xerogel process followed by drying and heat treatment at 200 °C for 30 min or at 550 °C for 1 h. XANES spectra were collected at the Sb L_1-edge at station E4 of HASYLAB and compared to reference powders Sb_2O_3 (for Sb^{3+}) and $FeSbO_4$ (for Sb^{5+}) to determine the oxidation state. EXAFS spectra were collected in transmission mode at the Sn *K*-edge and Sb *K*-edge on the ROMO II station at HASYLAB. XANES signals were fit using a linear combination of Sb^{3+} and Sb^{5+} dopants. The Sb^{3+}/Sb^{5+} ratios of all the samples at 200 °C were smaller than the ratios after treatment at 550 °C. For the 3, 4, 9, and 16% samples, the reported ratios were: 10/90 and 0/100, 15/85 and 7/93, 55/45 and 13/87, and 70/30 and 15/85. These results differed from the 1:3 ratios reported by Rockenberger *et al.* [64] for 9% and 17% samples; and the discrepancy was attributed to a rearrangement of antimony distributions arising from different preparation routes.

EXAFS results at the Sn *K*-edge indicated that for all the low-temperature samples, the intensity of the Sn–Sn/Sb shell was reduced, probably due to the small crystallite size, which ranged from 1.1(1) nm to 1.5(1) nm. After annealing at 550 °C, the grains for most doped samples grew to 2.0(2) to 2.8(4) nm while the intensities from the Sn–Sn/Sb shell got stronger. The weakest intensities were observed for the undoped and heavily doped samples, which maintained a small crystallite size (1.4(1) nm for undoped and 1.6(2) nm for 16% Sb-doped). The bond lengths for all samples were approximately the same, within experimental uncertainty: 2.06(1) Å, 3.21(1) Å, and 3.73(1) Å for the first, second, and third shells, respectively. For all samples and temperatures the coordination number for the first shell remained constant and ranged from 5.9(2) to 6.3(1). For the second shell, after the low temperature anneal, the number of neighbors ranged from 1.2(1) to 1.4(1). After the high temperature anneal, the numbers remained the same for the undoped and heavily doped samples, but it increased to up to 1.7(1) for the other samples. The third shell exhibited more differences. The coordination numbers ranged from 2.8(2) for undoped-SnO_2 to 3.6(2) for doped samples at low temperatures. At high temperatures, the values ranged from 3.4(2) to 5.2(3), with the undoped and heavily doped samples exhibiting the smallest numbers. The results indicated that for samples with ≤4 at.%, Sb was mostly in the Sb^{5+} oxidation state, and it favored crystallite growth resulting in stronger Sn- cation contributions that approached the behavior of bulk SnO_2.

The environment around Sb exhibited more dramatic differences among samples and thermal treatments. The intensities for the first and second shells continuously decreased with increasing doping after both annealing treatments; however as observed before, the intensity of the cation-cation shells got stronger after heating to 550 °C. The Sb-O bond length was 1.98(1) Å for all samples. The coordination number for this shell was 5.4(5) for the 1.5% Sb sample and continuously decreased with doping to 3.9(2) for the 16% Sb sample at 200 °C. This heavily doped sample looked similar to a Sb^{III}-grafted SnO_2 xerogel where the Sb^{3+}species are chemically adsorbed at the surface, and the coordination number is 3.1(3). At this low temperature, the authors proposed that antimony occupied superficial sites. At 550 °C, the corresponding coordination numbers ranged from 5.7(8) to 5.3(3), which approached the value of 6 for the Sn-O bond, therefore suggesting a more homogeneous incorporation of antimony in the lattice. The authors proposed that the Sb EXAFS data could be fit using a linear combination of grafted Sb^{3+} and Sb^{5+} species. For low doping levels below 4%, the treatment at 200 °C resulted in more than 85% of the Sb^{3+} species oxidizing to Sb^{5+}. As the doping increased, less than 50% of the Sb^{3+} species oxidized. At 550 °C, the Sb^{5+} species dominated and decreased with antimony content ranging from 100% to 85%.

3.6. F-Doped SnO_2

Fluorine-doped tin oxide (FTO) shows improved n-type conductivity and higher mobilities compared to pure TO and ATO [1]. FTO also exhibits good chemical stability making it suitable for applications in harsh environments. At low doping levels, the incorporation of fluorine results in increased conductivity, however at higher doping levels, the conductivity decreases. Theoretical calculations by Fantini and Torriani proposed that fluorine atoms occupy substitutional oxygen sites [66] and act as a donor species providing free carriers. Due to the weak x-ray scattering of fluorine and oxygen, such data could not be used to reliably obtain fractional occupancies of these light elements in the lattice. However XRD can be used to obtain lattice parameters and to study structural changes in a semi-quantitative manner. Several structural reports [67,68,69,70] have investigated the effect of fluorine doping tin oxide.

Canestrato *et al.* [67] prepared undoped, doped (FTO with 0.7% F) and heavily fluorine-doped SnO_2(hFTO with 1.3% F) thin films which were characterized using atomic force microscopy (AFM), XRF, and XRD. The electrical and optical properties of the samples were also measured and correlated to the doping content. The resistivity of the TO sample was $32(12) \times 10^{-4}\ \Omega$ cm. It decreased to $6(1) \times 10^{-4}\ \Omega$ cm for FTO, and increased to $25(8) \times 10^{-4}\ \Omega$ cm for hFTO, however the carrier concentration remained approximately the same ($\sim 6 \times 10^{20}$ cm^{-3}) for both doping levels. Doping decreased the transmittance of the films in the IR region due to conduction electrons, affecting more the hFTO samples. XRD measurements were conducted using a Cu K_α source. Based on the intensities of several diffraction peaks, the authors concluded that F occupied both interstitial (at 0, ½, ½) and substitutional sites for both FTO and hFTO; but at higher doping levels, the occupation of the interstitial site was enhanced. Figure 5 shows the location of these interstitial sites.

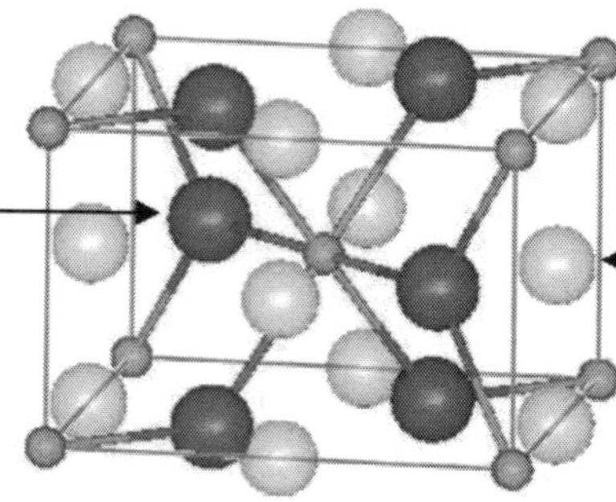

Figure 5. A unit cell of rutile SnO_2 showing the location of interstitial fluorine proposed by Canestrato *et al.* [67].

For hFTO, shifts in the Bragg angles were also consistent with the presence of interstitial atoms that caused slight distortion in the lattice. In the rutile structure, the interstitial fluorine species would act as acceptors leading to a decrease in n-type carriers. However while the conductivity decreased for hFTO, the carrier concentrations for the doped samples remained the same implying that the mobilities changed with doping. AFM showed that the morphologies for FTO and hFTO samples were comparable, ruling out the decrease in mobility due to grain boundary scattering. The authors concluded that mobility was reduced due to a combination of lattice distortion and carrier-impurity scattering, which increased with F concentration.

In a latter study, Canestrato *et al.* [68] developed a theoretical model for FTO that considered F on oxygen sites (F_O), F on interstitial sites (F_i) and neutralizing F_O-F_i defect complexes. The authors proposed that for low doping levels, F on oxygen sites created shallow donors that increased the conductivity, while F_i might be present but inactive. However for high doping levels, F_i species were present acting as compensating acceptors. Furthermore, in heavily doped FTO, the formation of F_O-F_i pairs was energetically favorable. In this complex defect, the acceptor role of F_i species resulted in a decrease in electrical conductivity.

Suffner *et al.* [69] studied the properties of nanocrystalline FTO using XRD, TEM, x-ray photoelectron spectroscopy (XPS), and Fourier-transformed infrared spectroscopy (FTIR). XRD patterns were collected using a Cu K_α source, and the data were analyzed using the Rietveld method [28] to determine particle size and cell parameters. The average particle size of the samples was approximately 7 nm from XRD, and 5 nm from TEM measurements. The doping concentrations determined by XPS were F/Sn = 0.109 and 0.271, which translated to F/O ratios of 1:15 and 1:6. The cell volumes obtained from XRD were 71.76 Å^3, 71.62 Å^3, and 71.70 Å^3 for the undoped TO, 0.109 and 0.271 FTO samples, respectively. The Sn/O ratio for pure TO was 1.80, and it decreased to 1.65 and then to 1.63 with increasing fluorine doping. These results supported the anion substitutional model, where fluorine gets incorporated on oxygen lattice sites. Because the Sn/O ratio did not change with increasing fluorine content and the volume of the cell increased, the authors proposed that at higher concentration, fluorine occupies an interstitial oxygen site. Theoretical calculations were performed to determine defect volumes of fluorine point defects and compare them with the XRD results. The calculated defects were F_O (fluorine on a lattice oxygen site), F_i (fluorine on an interstitial site) and an associate F_O-F_i defect ($F_{O,ap}$) that was proposed by Canestraro *et al.* [68]. Because the radius of fluorine is smaller compared to oxygen, it was expected that the lattice contracted with substitutional incorporation, as observed for the low doping sample. The model of isolated interstitial fluorine would lead to an overall increase in volume, while the associated defect configuration would lead to only a small cell expansion. Therefore, the authors concluded that at low doping levels, fluorine substituted for oxygen on a lattice site; and at high concentrations, the dopants went also into interstitial sites creating F_O-F_i defects.

A recent study by Chinnappa *et al.* [70] also investigated the effect of doping level on the structural and electrical properties of FTO films prepared via a spray technique. The films were prepared from a low concentration (0.1 M NH_4F, referred to as LC) and a high concentration (0.9 M NH_4F, referred to as HC) precursor, and the atomic percentage doping levels obtained were 0, 10, 20, 30, 40, and 50. XRD, energy-dispersive x-ray analysis (EDAX), and FTIR techniques were used to characterize the polycrystalline films. From the intensity of several Bragg reflections, the authors concluded that for LC heavily doped samples, substantial F_i species were present, while for HC films, the population of F_i species was not as pronounced even for high doping levels. These results correlated well with electrical measurements. For LC films, the conductivity and carrier concentration increased with increasing doping content but after 30% F, the samples became more resistive due to decreased electron concentration. This behavior could be explained by the increasing presence of F_i acceptor species. For HC samples, the carrier concentration and resistivity did not change much with doping level, therefore it seemed like the F_i population also remained unchanged. This suggested that not only doping level, but the preparation process had an effect on the electrical properties and the fluorine doping mechanism in rutile tin oxide.

3.7. Wurtzite Zinc Oxide (ZnO)

Zinc oxide has attracted much attention from the scientific community for several reasons: zinc is non-toxic, abundant, and less expensive than indium, making it an attractive substitute to indium-based TCOs. In the last ten years, recent efforts have also focused on achieving significant p-type doping [2,71,72,73,74,75], which is advantageous for commercial applications. Furthermore, ZnO exhibits green luminescence, which makes it even more desirable for industrial uses. Zinc oxide crystallizes in the hexagonal wurtzite structure with space group $P6_3mc$ (number 186) [76,77]. As shown in Figure 6, one unit cell has a total of two zinc and two oxygen ions. Zinc is tetrahedrally coordinated to oxygen. Three Zn-O bond lengths are 1.98 Å and the fourth length is 1.99 Å. Oxygen anions also exhibit tetrahedral coordination to zinc cations. The lattice parameters of pure ZnO at room temperature are $a = b = 3.2249$ Å, and $c = 5.2066$ Å, resulting in a c/a ratio of 1.602. The atomic lattice has enough space to accommodate interstitial defects.

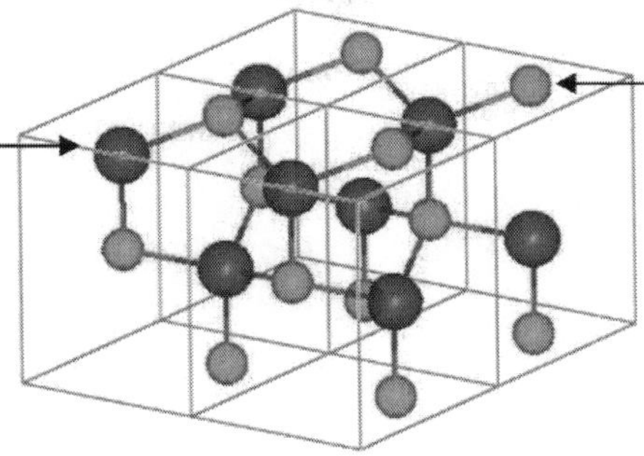

Figure 6. ZnO wurtzite structure (four unit cells are shown).

The defect structure of ZnO remains a controversial topic due to experimental difficulties in providing conclusive support for different proposed defect models. Theoretical calculations have aimed at explaining the n-type conductivity of undoped ZnO using various defect models [5,6,7,8,9,78]. Zinc interstitials (Zn_i), oxygen interstitials (O_i), oxygen vacancies (V_O), zinc vacancies (V_{Zn}), zinc antisites (Zn_O), oxygen antisites (O_{Zn}), hydrogen impurities, and/or defect complexes (such as V_O-Zn_i) have all been considered as possible defects. Formation energies for different charged, neutral, isolated, and complex defects and their location in the energy bands have been reported.

Kohan *et al.* [9] suggested that both zinc and oxygen vacancies were the most abundant native defects whose populations depend on the zinc partial pressure, and that the green luminescence was due to transitions between electrons in the conduction band and zinc vacancy levels and the presence of oxygen vacancies. On the other hand, Lany and Zunger [5] predicted that V_O was the major defect resulting in non-stoichiometry in equilibrium-grown ZnO, while Zn_i species had large formation energies. Van de Walle proposed that none of the native defects were responsible for the conductivity, but that the dominant defect consisted of H^+ impurities usually incorporated unintentionally during synthesis on interstitial sites [6]. Janotti and Van de Walle [78] concluded that zinc vacancies were deep acceptors with low formation energies that can act as compensating defects in n-type ZnO. Furthermore, their calculations indicated that oxygen vacancies were deep donors with high formation energies but donor impurities such as substitutional hydrogen (on an oxygen site) were most likely responsible for the n-type conductivity. Experimental reports [79,80,81,82] have pursued the idea of H as a cause of n-type conductivity in both intentionally hydrogen "doped" specimens and in samples without an external hydrogen source.

Lee *et al.* [83] proposed and calculated the formation energies of different types of oxygen interstitials. Erhart *et al.* [7,8] further investigated these defects that were predicted to exist under oxygen-rich conditions in the lower half of the band gap. The defects were referred to as octahedral interstitial ($O_{i,\,oct}$), and oxygen dumbbell interstitials (or split interstitial defects). The dumbbell configurations can be regular ($O_{i,\,db}$) or rotated ($O_{i,\,rot\text{-}db}$). The two oxygen atoms in the regular dumbbell defect had an oxidation state of −1 and occupied an oxygen lattice site. The defect was amphoteric acting as a donor and acceptor located at a deep state. Erhart *et al.* [7] proposed that regular dumbbell configuration may compensate for p-type doping in oxidizing conditions while the rotated dumbbell defect may compensate in n-type ZnO. Also under oxygen-rich conditions, but in the upper half of the band gap, zinc vacancies were predicted to be energetically stable. While under zinc-rich conditions, oxygen vacancies would be favored.

Ischenko *et al.* [84] prepared nanoparticles and studied their defects using XRD, EXAFS, FTIR, electron paramagnetic resonance (EPR), photoluminescence (PL), and magic-angle spinning nuclear magnetic resonance (MAS-NMR). The samples were prepared from an organometallic precursor. EXAFS measurements were conducted at the Hasylab E4 station at the Zn K edge in transmission mode. XRD measurements were performed using a conventional Cu K_α source and a position-sensitive detector. The diffraction data were analyzed to obtain lattice parameters and anisotropic particle size and strain. At 150 °C, XRD and EXAFS measurements indicated that long-range order was absent due to the presence of nanosized crystallites. As the temperature increased up to 450 °C, the particles grew to 25 nm and the XRD peaks sharpened. Coordination numbers and bond lengths were not reported but the EXAFS spectra from low thermolysis temperatures were significantly attenuated compared to bulk samples, indicating the effect of defects. As the temperature was increased, the EXAFS approached the bulk behavior. The *c* lattice parameter decreased almost linearly with annealing temperature, while the *a* lattice parameter exhibited a non-linear increase. This effect was consistent with the anisotropy in both the grain morphology and microstrain.

Ischenko *et al.* [84] also utilized EPR spectroscopy to study the presence of impurities and oxygen vacancies in the samples. Two types of paramagnetic signals were observed. A low-field signal was attributed to an unpaired electron trapped on an oxygen vacancy site (V_O^+), while the nature high-field signal was less clear. The authors proposed that the origin of the high-field signal was due to one electron weakly bound to ionized impurity atoms (e.g., C, Al, Ga, In). The authors also reported that under equilibrium conditions the predicted concentration of oxygen vacancies at 800 °C was extremely small, on the order of 10^{-5} mol% ($ZnO_{0.99993}$). By increasing the thermolysis rates at which samples were prepared, both signals became stronger indicating the presence of more defects under non-equilibrium conditions. Annealing at up to 800 °C resulted in a reduction of defect population. An interesting observation was that when ZnO was annealed in Zn vapor at 850 °C or in vacuum, the signal attributed to impurities was present but the signal due to oxygen vacancies was absent. The authors suggested that the effect of high Zn-rich conditions was the creation of more Zn_i defects instead of oxygen vacancies, contrary to theoretical predictions. Vlasenko and Watkins [85] created defects in ZnO by electron irradiation, and using optical detection of electron paramagnetic resonance (ODEPR) they detected the presence of Zn_i^+ in ZnO as an isolated species and as a Frenkel pair with V_{Zn}^-. These zinc defects were introduced at 4.2 K and were highly mobile with temperature annealing; however they disappeared by 200 K and were not stable at room temperature. This ODEPR study also found that oxygen vacancies were deep level defects [85,86]. The photoluminescence of the different samples prepared by Ischenko *et al.* [84] depended on sample preparation conditions, and no direct correlation was found between green luminescence and the concentration of oxygen vacancies as proposed by Kohan *et al.* [9].

EXAFS and XANES studies on thin films and nano-powders of ZnO were investigated by several research groups [87,88,89,90]. Tran *et al.* [87] studied nanocrystalline thin films grown under different water conditions. EXAFS data were collected at the Australian National Beamline Facility using a fluorescence detector. The films ranged from amorphous to crystalline and highly textured at higher water pressures. For the crystalline films, the particle size was calculated by examining the XRD peaks, and ranged from 15 nm to 40 nm. XPS data showed that up to 10 at.% of carbon was present and increased with decreasing water pressures. For all films the Zn-O and Zn-Zn distances were 1.97(5) Å and 3.25 Å, respectively, which agreed with the values for bulk ZnO. In ideal ZnO, the first coordination shell has 4 neighbors and the second has 12. When allowing the coordination numbers to vary, the analysis resulted in very low coordination numbers: 1.5 to 2 for the first shell and 2 to 5 for the second shell. The authors attributed these low values to the presence of oxygen vacancies. If the Debye-Waller factors were varied while keeping the coordination numbers fixed to 4 and 12, the disorder in the films was also evident from EXAFS results.

Chiou *et al.* [88] conducted angle-dependent XANES measurements at the O *K*, Zn L_3 and Zn *K* edges to study nanorods. The samples were 250 nm long and 45 nm in diameter. The authors concluded that the tip surfaces of the ZnO nanorods were terminated by oxygen ions and that the nanorods were oriented along the $[000\bar{1}]$ direction. Han *et al.* [89] also studied nanorods using orientation dependent EXAFS at the Zn *K* edge. The length of the rods along the *c* direction was 1 μm, while the diameters were 37(3) nm and 13(5) nm when grown at 500 °C and 800 °C, respectively. The *a* lattice parameters contracted by 0.04 Å while the *c* parameter expanded by 0.1 Å compared to bulk ZnO. The coordination numbers for the nanorods and the bond lengths for higher coordination shells were comparable to those of bulk. However the bond lengths for Zn-O differed from ideal ZnO. These authors concluded that the nanorods were terminated by oxygen, in agreement with the work of Chiou *et al.* [88].

Yu *et al.* [90] used angle-dependent XANES and EXAFS to study the local structure of ZnO thin films. XRD measurements were performed at the 10B X RS KIST-PAL beamline while EXAFS experiments were carried out at the 3 C1 beamline of the Pohang Light Source in Korea. The films were epitaxial, and strain decreased with increasing film thickness, up to 800 Å. The coordination numbers were fixed to ideal ZnO while bond lengths and Debye-Waller factors were refined. The first Zn-O coordination shells exhibited differences compared to bulk, while higher coordination shell bonds approached those of ideal ZnO. Also an increase in carrier concentration, Hall mobility and overall conductivity were measured as the films became thicker.

3.8. Aluminum-Doped Zinc Oxide (AZO)

Doped ZnO exhibits improved electrical properties compared to pure ZnO. Aluminum 3+ ions result in a higher electron concentration when they substitute for zinc 4+ ions. Electron concentrations on the order of 10^{20} cm^{-3} can be obtained [73]. The binary Al_2O_3-ZnO phase diagram prepared by Hansson *et al.* [91] determined that the equilibrium solubility of aluminum in ZnO expressed as mol fractions were 0.002 at 1250 °C, 0.005 at 1400 °C, 0.010 mol 1550 °C and 0.089 at 1695 °C. However, for thin films and nanopowders it is possible to reach much higher doping levels due to metastable synthesis conditions. The purpose of hydrogen anneals in n-type TCOs, as mentioned before, is to increase the conductivity of the materials. In the case of ZnO, as described in the previous section, the role of hydrogen as a dopant has been proposed. In AZO, several reports have also suggested that hydrogen plays a major role in defect creation and in enhancing the n-type conductivity [92].

Lany and Zunger [5] found that substitutional Al_{Zn} species in $ZnO_{0.99}Al_{0.01}O$ were shallow donors, which were expected to create free carrier due to their low formation energy. Theoretical calculations by Bazzani *et al.* [93] evaluated substitutional, interstitial and clusters of Al defects in AZO. For low doping levels, the Al-O bonds were on the order of 1.8 Å, close to the bond length in Al_2O_3. This length increased with doping. Their results proposed that up to 3% Al can be occupied in substitutional Zn sites. However, for higher doping concentrations, interstitial defects having large effective masses and low electron mobilities were expected to be significant and contribute to a decrease in conductivity. Also as the Al concentration increased, the Zn atoms were forced to move into interstitial positions changing the local coordination of the Zn-O bonds and resulting in inferior optical and electrical properties.

Brehm *et al.* [94] studied the local structure of Al-, Ga-, and In-doped nanocrystalline ZnO materials. XRD measurements were performed using a Cu K_α source and the data were analyzed using the Rietveld method [28] to obtain lattice parameters, microstrain, and grain size. EXAFS data were measured at the Zn, In, and Ga *K* edges in transmission mode at beamlines A1 and X1 at HASYLAB. EXAFS measurements at the Al *K* edge were not conducted but data from the Zn experiments were reported for AZO samples. EXAFS data were analyzed using a reverse Monte Carlo method that allows the simultaneous fit of spectra from two different absorbing elements [95]. The pair distribution functions obtained by this technique were analyzed to extract coordination numbers and distances. The AZO samples had 0, 2, and 5% Al with a particle size of 25.2(4), 13.3(6) nm and 11.1(4) nm, respectively. The ionic radius of Al^{3+} is smaller than that of Zn^{2+} for both tetrahedral and octahedral coordination. The *a* lattice parameter increased linearly with Al doping, while the *c* parameter decreased. Similar doping effects in particle size [96] and in the *c* parameter have been reported for nanopowders with Al-doping levels up to 25% [97]. The overall effect on doping was an increase in the volume of the lattice, a decrease in the *c*/*a* ratio, and an increase in microstrain probably due to the shortening of the Zn-O bond along the *c*-axis direction. Pair distribution function data for AZO at the Zn edge were modeled assuming substitutional Al and resulted in a splitting of the first Al peak. The authors interpreted this result as an indication that Al did not substitute into the Zn sites but rather occupied octahedral interstitial sites.

4. CONCLUSIONS

This manuscript has reviewed the defect structures of several indium-, tin-, and zinc-based n-type TCOs. Isolated and complex defects in these materials play an important role in the optical and electrical behavior of these materials. Understanding their defect mechanisms provides insight on the effect of doping, grain size, synthesis and treatment processes allowing an optimization of preparation techniques to meet the demands of commercial applications. X-ray and neutron scattering are powerful techniques that provide information about short- and long-range ordering of atoms, impurity determination and quantification, the location and population of most structural defects, and their effect on the microstructure of crystalline and amorphous TCOs. More experimental studies using these characterization techniques in combination with the measurement of electrical and optical properties and a close collaboration with theoreticians are needed to further elucidate the defect mechanisms of current and future TCOs.

ACKNOWLEDGMENTS

Funding for this project was provided by the Department of Physics at DePaul University.

REFERENCES

1. Chopra, K.L.; Major, S.; Pandya, D.K. Transparent conductors-a status review. Thin Solid Films 1988,102, 1–46.

2. Ginley, D.S.; Bright, C. Transparent conducting oxides. MRS Bulletin 2000, 25, 38–44.

3. Warschkow, O.; Ellis, D.E.; González, G.B.; Mason, T.O. Defect cluster aggregation and nonreducibility in tin-doped indium oxide. J. Am. Ceram. Soc. 2003, 86, 1707–1711.

4. Frank, G.; Köstlin, H. Electrical properties and defect model of tin-doped indium oxide layers. Appl. Phys. A Mat. Sci. Process. 1982, 27, 197–206.

5. Lany, S.; Zunger, A. Dopability, Intrinsic conductivity, and nonstoichiometry of transparent conducting oxides. Phys. Rev. Lett. 2007, 98, 045501:1–045501:4.

6. Van de Valle, C.G. Hydrogen as a cause of doping in zinc oxide. Phys. Rev. Lett. 2000, 85, 1012–1015.

7. Erhart, P.; Klein, A.; Albe, K. First-principles study of the structure and stability of oxygen defects in zinc oxide. Phys. Rev. B 2005, 72, 085213:1–085213:7.

8. Erhart, P.; Klein, A.; Albe, K. First-principles study of intrinsic point defects in ZnO: Role of band structure, volume relaxation, and finite-size effects. Phys. Rev. B 2006, 73, 205203:1–205203:9.

9. Kohan, A.F.; Ceder, G.; Morgan, D.; van de Walle, C.G. First-principles study of native point defects in ZnO. Phys. Rev. B 2000, 61, 15019–15027.

10. Warren, B.E. The diffraction of x-rays in glass. Phys. Rev. 1934, 45, 657–661.

11. Klug, H.P.; Alexander, L.E. X-ray Diffraction Procedures for Polycrystalline and Amorphous Materials, 2nd ed.; John Wiley & Sons: New York, NY, USA, 1974; pp. 1–996.

12. Petkov, V.; Shastri, S.D. Element-specific structure of materials with intrinsic disorder by high-energy resonant x-ray diffraction and differential atomic pair-distribution functions: A study of PtPd nanosized catalysts. Phys. Rev. B 2010, 81, 165428:1–165428:8.

13. Petkov, V.; Billinge, S.J.L.; Shastri, S.D.; Himmel, B. Polyhedral units and network connectivity in calcium aluminosilicate glasses from high-energy x-ray diffraction. Phys. Rev. Lett. 2010, 85, 3436–3439.

14. Marezio, M. Refinement of the crystal structure of In2O3 at two wavelengths. Acta Crystallogr. 1966, 20, 723–728.

15. Prince, E. The International Union of Crystallography, International Tables for Crystallography; Wilson, A.J.C., Ed.; Kluwer Academic Publishers: Dordrecht, The Netherlands, 1992.

16. De Wit, J.H.W. Structural aspects and defect chemistry in In2O3. J. Solid State Chem. 1973, 8, 142–149.

17. De Wit, J.H.W. The high temperature behavior of In2O3. J. Solid State Chem. 1975, 13, 192–200.

18. De Wit, J.H.W. Electrical properties of In2O3. J. Solid State Chem. 1977, 20, 143–148.

19. De Wit, J.H.W.; van Unen, G.; Lahey, M. Electron concentration and mobility in In2O3. J. Solid State Chem. 1977, 38, 819–824.

20. González, G.B.; Okasinski, J.S.; Mason, T.O.; Buslaps, T.; Honkimäki, V. In situ studies on the kinetics of formation and crystal structure of In4Sn3O12 using high-energy x-ray diffraction. J. Appl. Phys. 2008,104, 043520:1–043520:8.

21. González, G.B.; Mason, T.O.; Okasinski, J.S.; Buslaps, T.; Honkimäki, V. Determination of the solubility of tin in indium oxide using in situ and ex situ x-ray diffraction. J. Am. Ceram. Soc. 2012, 95, 809–815.

22. González, G.B. Studies on the Defect Structure of Indium-Tin Oxide Using X-ray and Neutron Diffraction. Ph.D. Dissertation, Northwestern University, Evanston, IL, USA, 2003.

23. González, G.B.; Mason, T.O.; Quintana, J.P.; Warschkow, O.; Ellis, D.E.; Hwang, J.H.; Hodges, J.P. Defect structure studies of bulk and nano-indium-tin oxide. J. Appl. Phys. 2004, 96, 3912–3920.

24. González, G.B.; Cohen, J.B.; Hwang, J.-H.; Mason, T.O.; Hodges, J.P.; Jorgensen, J.D. Neutron diffraction study on the defect structure of indium-tin oxide. J. Appl. Phys. 2001, 89, 2550–2555.

25. Parent, H.; Dexpert, H.; Tourillon, G.; Grimal, J.-M. Structural study of tin-doped indium oxide thin films using x-ray absorption spectroscopy and x-ray diffraction I. Description of the indium site. J. Electrochem. Soc. 1992, 139, 276–281.

26. Parent, H.; Dexpert, H.; Tourillon, G.; Grimal, J.-M. Structural study of tin-doped indium oxide thin films using x-ray absorption spectroscopy and x-ray diffraction II. Tin environment. J. Electrochem. Soc. 1992, 139, 282–285.

27. Nadaud, N.; Lequeux, N.; Nanot, M.; Jové, J.; Roisnel, T. Structural studies of tin-doped indium oxide (ITO) and In4Sn3O12. J. Solid State Chem. 1998, 135, 140–148.

28. Rietveld, H.M. A profile refinement method for nuclear and magnetic structure. J. Appl. Crystallogr.1969, 2, 65–71.

29. Warschkow, O.; Ellis, D.E.; González, G.B.; Mason, T.O. Defect structures of tin-doped indium oxide. J. Am. Ceram. Soc. 2003, 86, 1700–1706.

30. Warschkow, O.; Miljacic, L.; Ellis, D.E.; González, G.B.; Mason, T.O. Interstitial oxygen in tin-doped indium oxide transparent conductors. J. Am. Ceram. Soc. 2006, 89, 616–619.

31. Inerbaev, T.M.; Sahara, R.; Mizuseki, H.; Kawazoe, Y.; Nakamura, T. Interstitial oxygen and dopant atoms arrangement in tin-doped indium oxide. Mater. Trans. 2007, 48, 666–669.

32. Inerbaev, T.M.; Sahara, R.; Mizuseki, H.; Kawazoe, Y.; Nakamura, T. Reducible and non-reducible defect clusters in tin-doped indium oxide. Solid State Comm. 2010, 150, 18–21.

33. Phillips, J.M.; Cava, R.J.; Thomas, G.A.; Carter, S.A.; Kwo, J.; Siegrist, T.; Krajewski, J.J.; Marshall, J.H.; Peck, W.F.; Rapkine, D.H. Zinc-indium-oxide: A high conductivity transparent conducting oxide.Appl. Phys. Lett. 1995, 67, 115–118.

34. Palmer, G.B.; Poeppelmeier, K.R.; Mason, T.O. Zn2−xSn1−xIn2xO4−δ: An indium-substituted spinel with transparent conducting properties. J. Solid State Chem. 1997, 134, 192–197.

35. Palmer, G.B.; Poeppelmeier, K.R.; Mason, T.O. Conductivity and transparency of ZnO/SnO2-cosubstituted In2O3. Chem. Mater. 1997, 9, 3121–3126.

36. Palmer, G.B. Phase Relations and Physical Properties in the (Indium, Gallium)-Tin-Zinc Oxide Systems. Ph.D. Dissertation, Northwestern University, Evanston, IL, USA, 1999.

37. Ambrosini, A.; Malo, S.; Poeppelmeier, K.R.; Lane, M.A.; Kannewurf, C.R.; Mason, T.O. Zinc codoping in cosubstituted In2−2xSnxZnxO3−δ. Chem. Mater. 2002, 14, 58–63.

38. Bizo, L.; Choisnet, J.; Retoux, R.; Raveau, B. The great potential of coupled substitutuions in In2O3 for the generation of bixbyite-type transparent conducting oxides, In2−2xMxSnxO3. Solid State Commun.2005, 136, 163–168.

39. Seo, K.H.; Park, D.H.; Lee, J.H.; Kim, J.J. Co-doping effect of SnO2 and ZnO in In2O3 ceramics: Change in solubility limit and electrical properties. J. Solid State Ionics 2006, 177, 601–605.

40. Harvey, S.P.; Mason, T.O.; Buchholz, D.B.; Chang, R.P.H.; Körber, C.; Klein, A. Carrier generation and inherent off-stoichiometry in Zn, Sn codoped indium oxide (ZITO) bulk and thin-film specimens. J. Am. Ceram. Soc. 2008, 91, 467–472.

41. Harvey, S.P.; Poeppelmeier, K.R.; Mason, T.O. Subsolidous phase relationships in the ZnO-In2O3-SnO2 system. J. Am. Ceram. Soc. 2008, 91, 3683–3689.

42. Harvey, S.P.; Mason, T.O.; Körber, C.; Klein, A. Bulk defect chemistry and surface electronic behavior of Zn, Sn codoped In2O3 transparent conducting oxides. Phys. Chem. Phys. 2009, 11, 3099–3104.

43. Hoel, C.A.; Gaillard, J.-F.; Poeppelmeier, K.R. Probing the local structure of crystalline ZITO: In2−2xSnxZnxO3 (x ≤ 0.4). J. Solid State Chem. 2010, 183, 761–768.

44. Hoel, C.A.; Amores, J.M.G.; Moran, E.; Alario-Franco, M.A.; Gaillard, J.-F.; Poeppelmeier, K.R. High-pressure synthesis and local structure of corundum-type In2−2xZnxO3 (x ≤ 0.7). J. Am. Chem. Soc. 2010,132, 16479–16487.

45. Hoel, C.A.; Mason, T.O.; Gaillard, J.-F.; Poeppelmeier, K.R. Transparent conducting oxides in the ZnO-In2O3-SnO2 system. Chem. Mater. 2010, 22, 3569–3579.

46. Hoel, C.A.; Xie, S.; Benmore, C.; Malliakas, C.D.; Gaillard, J.-F.; Poeppelmeier, K.R. Evidence for Tetrahedral Zinc in Amorphous

In2−2xZnxSnxO3 (a-ZITO). J. Inorg. Gen. Chem. 2011, 637, 885–894.

47. Proffit, D.E.; Buchholz, D.B.; Chang, R.P.H.; Bedzyk, M.J.; Mason, T.O.; Ma, Q. X-ray absorption spectroscopy study of the local structures of crystalline Zn-In-Sn oxide thin films. J. Appl. Phys. 2009,106, 113524:1–113524:6.

48. Proffit, D.E. Structure and Electronic Properties of Crystalline and Amorphous Zn-In-Sn-O Thin Films. Ph.D. Dissertation, Northwestern University, Evanston, IL, USA, 2011.

49. Buchholz, D.B.; Liu, J.; Marks, T.J.; Zhang, M.; Chang, R.P.H. Control and characterization of the structural, electrical, and optical properties of amorphous zinc-indium-tin oxide thin films. ACS Appl. Mater. Interfaces 2009, 1, 2147–2153.

50. Adurodija, F.O.; Semple, L.; Brüning, R. Crystallization process and electro-optical properties of In2O3and ITO thin films. J. Mater. Sci. 2006, 41, 7096–7102.

51. Adurodija, F.O.; Semple, L.; Brüning, R. Real-time in situ crystallization and electrical properties of pulsed laser deposited indium oxide thin films. Thin Solid Films 2005, 492, 153–157.

52. Gordon, R.G. Criteria for choosing transparent conductors. MRS Bulletin 2000, 25, 52–57.

53. Baur, W.H. Über die Verfeinerung der Kristallstrukturbestimmung einiger Vertreter des Rutiltyps: TiO2, SnO2, GeO2 und MgF2. Acta Crystallogr. A 1956, 9, 515–520.

54. Samson, S.; Fonstad, C.G. Defect structure and electronic donor levels in stannic oxide crystals. J. Appl. Phys. 1973, 44, 4618–4621.

55. Davis, S.R.; Chadwick, A.V.; Wright, J.D. A combined EXAFS and diffraction study of pure and doped nanocrystalline tin oxide. J. Phys. Chem. B 1997, 101, 9901–9908.

56. Savin, S.L.P.; Chadwick, A.V.; O'Dell, L.A.; Smith, M.E. Structural studies of nanocrystalline oxides.Solid State Ionics 2006, 177, 2519–2526.

57. Chadwick, A.V. X-ray absorption studies of the structure of nanocrystalline oxides. Solid State Ionics2006, 177, 2481–2485.

58. Berry, F.J.; Laundry, B.J. Antimony-121 Mössbauer study of the effects of calcinations on the structure of tin-antimony oxides. J. Chem. Soc. Dalton. 1981, 1442–1444.

59. Kojima, M.; Kato, H.; Gatto, M. Blackening of tin oxide thin-films heavily-doped with antimony.Philos. Mag. B 1993, 68, 215–222.

60. Pyke, D.R.; Reid, R.; Tilley, R.J.D. Structures of tin oxide-antimony oxide catalysts. J. Chem. Soc. Faraday Trans. 1 1980, 76, 1174–1182.

61. Gržeta, B.; Tkalčec, E.; Goebbert, C.; Takeda, M.; Takahashi, M.; Nomura, K.; Jakšić, M. Structural studies of nanocrystalline SnO2 doped with antimony: XRD and Mossbauer spectroscopy. J. Phys. Chem. Solids 2002, 63, 765–772.

62. Montilla, F.; Morallón, E.; Battisti, A.; Benedetii, A.; Yamashita, H.; Vázquez, J.L. Preparation and characterization of antimony-doped tin dioxide electrodes. Part 2. XRD and EXAFS characterization. J. Phys. Chem. B 2004, 108, 5044–5050.

63. Berry, F.J.; Greaves, C. A neutron diffraction investigation of the defect rutile structure of tin-antimony oxide. J. Chem. Soc. Dalton. 1981, 2447–2451.

64. Rockenberger, J.; Felde, U.; Tischer, M.; Tröger, L.; Haase, M.; Weller, H. Near edge x-ray absorption fine structure measurements (XANES) and extended x-ray absorption fine structure measurements (EXAFS) of the valence state and coordination of antimony in doped nanocrsytalline SnO2. J. Chem. Phys. 2000, 112, 4296–4304.

65. Geraldo, V.; Briois, V.; Scalvi, L.V.A.; Santilli, C.V. Structural characterization of nanocrystalline Sb-doped SnO2 xerogels by

multiedge x-ray absorption spectroscopy. J. Phys. Chem. C 2010, 114, 19206–19213.

66. Fantini, M.; Torriani, I. The compositional and structural properties of sprayed SnO2:F thin films. Thin Solid Films 1986, 138, 255–265.

67. Canestraro, C.D.; Oliveira, M.M.; Valaski, R.; da Silva, M.V.S.; David, D.G.F.; Pepe, I.; da Silva, A.F.; Roman, L.S.; Persson, C. Strong inter-conduction-band absorption in heavily fluorine doped tin oxide.Appl. Surf. Sci. 2008, 225, 1874–1879.

68. Canestraro, C.D.; Roman, L.S.; Persson, C. Polarization dependence of the optical response in SnO2and the effects from heavily F doping. Thin Solid Films 2009, 517, 6301–6304.

69. Suffner, J.; Agoston, P.; Kling, J.; Hahn, H. Chemical vapor synthesis of fluorine-doped SnO2 (FTO) nanoparticles. J. Nanopart. Res. 2010, 12, 2579–2588.

70. Chinnappa, L.; Ravichandran, K.; Saravanakumar, K.; Muruganantham, G.; Sakthivel, B. The combined effects of molar concentration of the precursor solution and fluorine doping on the structural and electrical properties of tin oxide films. J. Mater. Sci. Mater. El. 2011, 22, 1827–1834.

71. Look, D.C. Recent advances in ZnO materials and devices. Mater. Sci. Eng. B 2001, 80, 383–387.

72. Look, D.C.; Claflin, B. P-type doping and devices based on ZnO. Phys. Stat. Sol. B 2004, 241, 624–630.

73. Klingshirn, C. ZnO: From basics towards applications. Phys. Stat. Sol. B 2007, 244, 3027–3073.

74. Li, X.; Yan, Y.; Gessert, T.A.; Perkins, C.L.; Young, D.; DeHart, C.; Young, M.; Coutts, T.J. Chemical vapor deposition-formed p-type ZnO thin films. J. Vac. Sci. Technol. A 2003, 21, 1342–1346.

75. Transparent Conductive Zinc Oxide: Basics and Applications in Thin Film Solar Cells; Ellmer, K., Klein, A., Rech, B., Eds.; Springer

Series in Materials Science; Springer-Verlag: Berlin, Germany, 2008.

76. Bragg, W.L. The crystalline structure of zinc oxide. Phil. Mag. 1920, 39, 647–651.

77. Abrahams, S.C.; Bernstein, J.L. Remeasurement of the structure of hexagonal ZnO. Acta Crystallogr. B1969, 25, 1233–1236.

78. Janotti, A.; van de Walle, C.G. Native point defects in ZnO. Phys. Rev. B 2007, 76, 165202:1–165202:22.

79. Cox, S.F.J.; Davis, E.A.; Cottrell, S.P.; King, P.J.C.; Lord, J.S.; Gil, J.M.; Alberto, H.V.; Vilao, R.C.; Duarte, J.P.; de Campos, N.A.; et al. Experimental confirmation of the predicted shallow donor hydrogen state in zinc oxide. Phys. Rev. Lett. 2001, 86, 2601–2604.

80. Shi, G.A.; Saboktakin, M.; Stavola, M.; Pearton, S.J. "Hidden" hydrogen in as-grown ZnO. Appl. Phys. Lett. 2004, 85, 5601–5603.

81. Chen, L.Y.; Chen, W.H.; Wang, J.J.; Hong, F.C.N.; Su, Y.K. Hydrogen-doped high conductivity ZnO films deposited by radio-frequency magnetron sputtering. Appl. Phys. Lett. 2004, 85, 5628–5630.

82. Zhu, B.L.; Wang, J.; Zhu, S.J.; Wu, J.; Zeng, D.W.; Xie, C.S. Influence of hydrogen introduction on structure and properties of ZnO thin films during sputtering and post-annealing. Thin Solid Films2011, 519, 3809–3815.

83. Lee, E.-C.; Kim, Y.-S.; Jin, Y.-G.; Chang, K.J. Compensation mechanism for N acceptors in ZnO. Phys. Rev. B 2001, 64, 085120:1–085120:5.

84. Ischenko, V.; Polarz, S.; Grote, D.; Stavarache, V.; Fink, K.; Driess, M. Zinc oxide nanoparticles with defects. Adv. Funct. Mater. 2005, 15, 1945–1954.

85. Vlasenko, L.S.; Watkins, G.D. Intrinsic defects in ZnO: A study using optical detection of electron paramagnetic resonance. Physica B 2006, 376, 677–681.

86. Stehr, J.E.; Meyer, B.K.; Hofmann, D.M. Magnetic resonance of impurities, intrinsic defects and dopants in ZnO. Appl. Magn. Reson. 2010, 39, 137–150.

87. Tran, N.H.; Hartmann, A.J.; Lamb, R.N. Structural order of nanocrystalline ZnO films. J. Phys. Chem. B1999, 103, 4264–4268.

88. Chiou, J.W.; Jan, J.C.; Tsai, H.M.; Bao, C.W.; Pong, W.F.; Tsai, H.-H.; Hong, I.-H.; Klauser, R.; lee, J.F.; Wu, J.J.; Liu, S.C. Electronic structure of ZnO nanorods studied by angle-dependent x-ray absorption spectroscopy and scanning photoelectron microscopy. Appl. Phys. Lett. 2004, 84, 3462–3465.

89. Han, S.-W.; Yoo, H.-J.; An, S.J.; Yoo, J.; Yi, G.-C. Orientation-dependent x-ray absorption fine structure of ZnO nanorods. Appl. Phys. Lett. 2005, 86, 021917:1–021917:4.

90. Yu, C.J.; Sung, N.E.; Lee, H.K.; Shin, H.J.; Yun, Y.D.; Kang, S.W.; Lee, I.J. Structural properties of low-temperature grown ZnO thin films determined by x-ray diffraction and x-ray absorption spectroscopy. Thin Solid Films 2011, 519, 4366–4370.

91. Hansson, R.; Haynes, P.C.; Jak, E. Experimental study of phase equilibria in the Al-Fe-Zn-O system in air. Metall. Mater. Trans. B 2004, 35B, 633–642.

92. Chang, H.P.; Wang, F.H.; Wu, J.Y.; Kung, C.Y.; Liu, H.W. Enhanced conductivity of aluminum doped ZnO films by hydrogen plasma treatment. Thin Solid Films 2010, 518, 7445–7449.

93. Bazzani, M.; Neroni, A.; Calzolari, A.; Catellani, A. Optoelectronic properties of Al:ZnO: Critical dosage for an optimal transparent conductive oxide. Appl. Phys. Lett. 2011, 98, 121907:1–121907:4.

94. Brehm, J.U.; Winterer, M.; Hahn, H. Synthesis and local structure of doped nanocrsytalline zinc oxides. J. Appl. Phys. 2006, 100, 064311:1–064311:9.

95. Winterer, M. Reverse Monte Carlo analysis of extended x-ray absorption fine structure spectra of monoclinic and amorphous zirconia. J. Appl. Phys. 2000, 88, 5635–5644.

96. Zhang, Y.; Yang, Y.; Zhao, J.; Tan, R.; Wang, W.; Cui, P.; Song, W. Optical and electrical properties of aluminum-doped zinc oxide nanoparticles. J. Mater. Sci. 2011, 46, 774–780.

97. Kumar, R.S.; Sathyamoorthy, R.; Sudhagar, P.; Matheswaran, P.; Hrundhya, C.P.; Kang, Y.S. Effect of aluminum doping on the structural and luminescent properties of ZnO nanoparticles synthesized by wet chemical method. Physica E 2011, 43, 1166–1170.

6

SMALL ANGLE NEUTRON SCATTERING AND X-RAY DIFFRACTION STUDIES OF NANOCRYSTALLINE TITANIUM DIOXIDE

Muhammad Nasir Khan, Javaid Bashir

Physics Division, PINSTECH, Islamabad, Pakistan

ABSTRACT

Nanocrystalline titanium dioxide powder is characterized for phase analysis as well as particle size and its distribution by x-ray diffraction and small angle neutron scattering measurements. Analysis of the SANS data in the momentum transfer range $q = 0.1 - 1.8\ nm^{-1}$ reveals an average particle size of 24.82 nm in good agreement with the particle size determined earlier by transmission electron microscopy. XRD measurement proves co-existence of rutile and anatase phases in this commercial TiO_2 nanocrystalline powder.

Keywords: Titanium Dioxide, Average Size, SANS, Microstructure, X-Ray Diffraction

1. INTRODUCTION

Microstructural aspects such as particle size and its distribution, defect structures of the materials having high technological applications are of great interest from both safety and performance perspectives. Nanophase oxide materials are interesting due to their potential and unique mechanical, chemical, electrical, optical and magnetic properties. Among various dioxides titanium dioxide (TiO_2) has been suggested as being a material that could have a series of phases with hardness possibly approaching diamond [1,2]. Pure as well as doped nanostructured TiO_2 materials are essential for the development of advanced electrochemical solar cells, photo catalytic, electron sensitive devices and optical coatings. TiO_2 in its entire known three phases is technologically important material for numerous applications. TiO_2 with rutile phase has been widely used as white pigment material because of its good scattering effect that protects materials from ultraviolet light. It has also been used for optical coatings, beam splitters because of its high dielectric constant and refractive index. The material also possesses good oil adsorption ability, tinting power, and chemical stability under strong acidic or basic conditions [3,4]. TiO_2 with an anatase structure has been used as a photocatalyst for photodecomposition and solar energy conversion because of its high photoactivity [5-8]. TiO_2 shows different electrical characteristics with oxygen partial pressure, because it has wide chemical stability and a non-stoichiometric phase region. For this reason, it can also be used as humidity sensor and high temperature oxygen sensors [9,10].

Nanocrystalline TiO_2 powders are generally synthesized by different techniques such as the sulphate or chloride processes, sol-gel method, and hydrothermal synthesis [11-13]. All these synthesis techniques are vital and have advantages/disadvantages over each other but still there is a need for simple and cost effective process for the commercial production of TiO_2 nano powders. For the practical applications of these high technological materials their physical and chemical properties are greatly dependent on the particle size. Therefore, it is necessary to know the true particle size, shape, and their distribution in these nanocrystalline materials. X-ray diffraction (XRD) is a powerful tool to identify the different phases of compound but small angle neutron scattering (SANS) is well suited technique for an analysis of microstructural characterization of such high technological materials, allowing for quantitative measurements of particle/pores morphology, particle size distribution. In this paper the results are demonstrated for the nanocrystalline commercial DEGUSSA P25 TiO_2 for particle size determination and distribution by SANS and phase analysis by XRD. The results are being compared with earlier published data.

2. EXPERIMENTAL

The XRD pattern for TiO_2 nanocrystalline powder was measured by Philips Xpert PRO diffractometer equipped with PW3015 generator, goniometer and X'celerator (SSD detectors) in the 2q ranges 20° - 90°. The small angle neutron scattering measurement was carried out with the SANS instrument at the HANARO reactor of the Korea Atomic Energy Research Institute, Korea. The SANS instrument had a typical pinhole geometry using neutrons with wavelength of 5.08 Å. Wavelength spread was (Dl/l » 10%). The distance from sample to detector was 3 m. The measured q-value range was 0.1 - 1.8 nm^{-1}. The scattering pattern of the sample was recorded with a two-dimensional position sensitive detector with an active area of 64.5 ´ 64.5 cm^2, and were circularly averaged to produced one-dimensional intensity profiles. The scattering intensity data were reduced to absolute scale by using standard porous silica in 1 mm path length quartz cell. Scattering intensities were measured from TiO_2 nanocrystalline powder filled in 25 mm diameter quartz cell with 1 mm path length.

3. RESULTS & DISCUSSION

Figure 1 shows the XRD pattern for the nanocrystalline TiO_2 powder. All of the peaks on this pattern belong to rutile and anatase phases of this compound. These peaks have been marked as (A) for anatase and (R) for rutile phase respectively on the diffraction pattern. The general proportion of these phases in commercial TiO_2 has been known as 75% for anatase and 25% for rutile [12]. In **Figure 1**, the XRD data gives this ratio as 75%:25% for anatase and rutile phases on the basis of maximum peak intensities ratios for the respective two phases. No traces of others TiO_2 phases such as brookite or Zr based srilankite were observed which often appear in most of the nanocrystalline TiO_2 powders prepared by different processes [12]. The particle size in the rutile phase was reported to be larger than that of anatase by different synthesis processes because of the agglomeration process, which is the growth of primary TiO_2particles with an increase of thermal energy [12]. The above finding of the phases and their ratio in this study are well in agreement with the earlier published X-ray and synchrotron X-ray diffraction data on the same material [14,15].

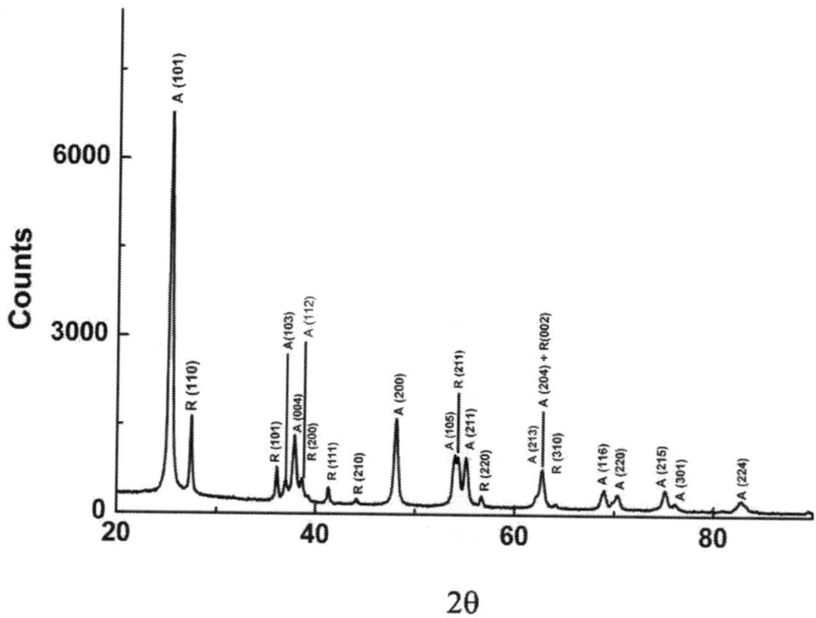

Figure 1. XRD pattern for the Degussa P25 nanocrystalline TiO_2 powder.

Figure 2 shows the one-dimensional (1-D) SANS pattern for the TiO_2 nanocrystalline powder circularly averaged from the 2-D measured pattern. The 2-D pattern (not shown here) clearly revealed isotropic scattered symmetric behavior, thus indicating that structures inside the powders are spherical. As the particles in the powder possess different sizes and internal structures, therefore the macroscopic differential scattering cross section for spherical particles as reported in [16] is given by;

$$\frac{d\Sigma}{d\Omega}(Q)=\Delta\eta^2\int_0^\infty\left(\frac{4}{3}\pi R^3\right)^2 N(R)F^2(Q,R)\,dR \tag{1}$$

where $\Delta\eta^2$ is the scattering contrast (η is the scattering length density) depending on the chemical composition of both the scattering centers and matrix, R is the radius of the spherical particles. N(R)dR is the number of the particles per unit volume with a typical size between R and R + dR. $F^2(Q, R)$ is the form factor which in the case of spherical particles is given by;

$$F(Q,R)=3\frac{\sin(QR)-QR\cos(QR)}{(QR)^3} \tag{2}$$

The scattering length density was calculated with assumption that the particles smaller than 50 nm are dispersed equally in this material. The model fitting of the real space size distribution, log-normal size distribution to the scattering pattern was performed using non linear least square fitting [17];

$$N(R)=\sum_i N_{oi}\exp\left[-0.5\left(\frac{\ln(R)-\ln(R_{oi})}{S_i}\right)^2\right] \tag{3}$$

And

$$\langle R\rangle=\frac{\int RN(R)\,dR}{\int N(R)\,dR} \tag{4}$$

where N_o, R_o and S are the standard scaling factors, center and width parameters of this distribution type. The lognormal distribution was fitted in the q-range 0.1 - 1.1 nm^{-1} in Equation (1). The results of the fitting along with the observed data are shown in **Figure 2**.

The average $\langle R\rangle$, of the particles calculated from the fitted parameters comes out to be 9.62 nm for the nanocrystalline commercial DEGUSSA P25 material as shown in **Figure 3** from the number distribution function curve determined using the spherical model calculations

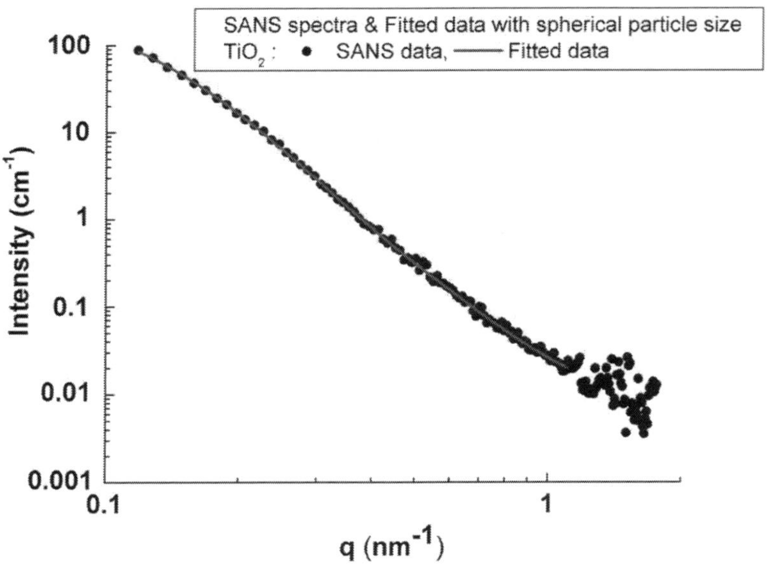

Figure 2. 1-D SANS Pattern averaged from the 2-D pattern for the nanocrystalline TiO_2 powder sample; (solid lines shows spherical model fitting to the observed data).

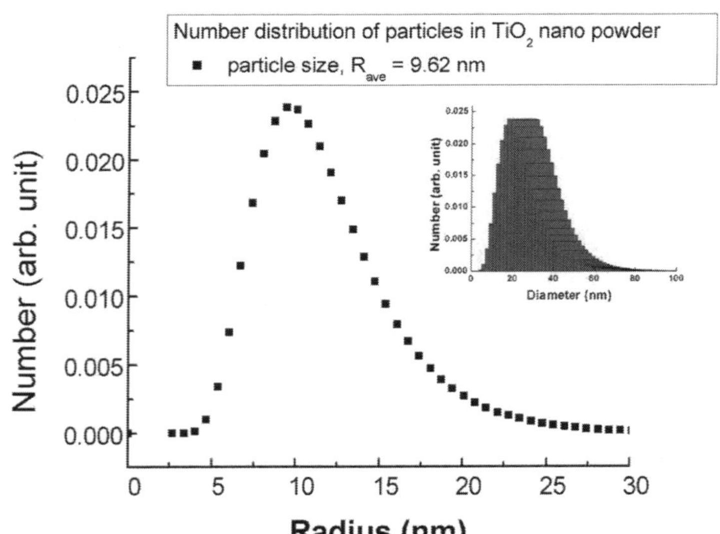

Figure 3. Particle size distribution for the nanocrystalline TiO_2 powder sample (Inset shows the true particle diameter distribution).

mentioned above. Similar approached was used to estimate the size and microstructure of precipitates affected by coiling temperature in ultra low carbon steel in [18].

The size distribution was calculated in terms of the true particles diameter by the formula given below as mentioned in [19] and is shown in the inset of **Figure 3**.

$$D = (5/3)^{\frac{1}{2}} 2R_g \tag{5}$$

The average diameter of the spherical particle comes out to be 24.82 nm from SANS results, which corresponds well with the particles size determined by SunJae Kim, et al. [20] using transmission electron microscopy technique.

4. CONCLUSIONS

For nanocrystalline DEGUSSA P25, TiO_2 powder the SANS results revealed that the average particle size of 24.82 nm agrees well with the earlier results of TEM. The XRD pattern proved coexistence of anatase and rutile phases with ratio of 75% and 25% respectively. Furthermore, the results demonstrates the power and utility of the SANS technique in characterizing microstructure, particle size, and distribution in nanocrystalline materials.

5. ACKNOWLEDGEMENTS

The authors are thankful to Mr. Chang-Hee Lee, B. S. Seong and Shin Enjoo of Neutron Scattering Division, KAERI, for their help in the SANS experiment and and data analysis. Their local hospitality is also highly acknowledged.

REFERENCES

1. J. M. Leger, J. Haines and B. Blanzat, "Materials Potentially Harder than Diamond: Quenchable High—Pressure Phases of Transition Metal Oxides," Journal of Materials Science Letters, Vol. 13, 1994, pp. 1688-1690.

2. A. Fahmi, C. Minot, B. Silvi and M. Causa, "Theoretical Analysis of the Structures of TiO_2Crystals," Physical Review B, Vol. 47, 1993, pp. 11717-11724.

3. H. Cheng, J. Ma, Z. Zhao and L. Qi, "Hydrothermal Preparation of Uniform Nanosize Rutile and Anatase Particles," Chemistry of Materials, Vol. 7, No. 4, 1995, pp. 663-671.

4. R. U. Flood and D. Fitzmaurice, "Preparation, Characterization and Potential Dependent Optical Absorption Spectroscopy of Unsupported Large Area Transparent Nanocrystalline TiO_2 Membranes," Journal of the Physical Chemistry, Vol. 99, No. 22, 1995, pp. 8954-8958.

5. T. Fuyuki and H. Matsunami, "Electronic Properties of the Interface between Si and Tio_2Deposited at Very Low Temperatures," Japanese Journal of Applied Physics, Vol. 25, No. 9, 1986, pp. 1288-1291.

6. P. V. Kamat and N. M. Dimitrijevic, "Colloidal Semiconductors as Photo-Catalyst for Solar Energy Conversion," Solar Energy, Vol. 44, 1990, pp. 83-98.

7. S. A. Larson and J. L. Falconer, "Characterization of TiO_2 Photocatalysis Used in Trichlorethene Oxidation," Applied Catalysis B: Environmental, Vol. 4, 1994, pp. 325-342.

8. A. L. Micheli, "Fabrication and Performance Evaluation of Titania Automotive Exhaust Gas Sensor," American Ceramics Society Bulletin, Vol. 54, 1984, pp. 694-698.

9. K. L. Siefering and G. L. Griffin, "Kinetics of Low Pressure Chemical Vapor Deposition $TiO_2$from Titanium Tetraisopropxide," Journal of the Electrochemical Society, Vol. 137, No. 4, 1990, pp. 814-818.

10. H. Tang, K. Prasad, R. Sanjines and F. Levy, "TiO_2 Anatase Thin Films as Gas Sensors," Sensors Actuators B, Vol. 26-27, 1995, pp. 71-75.

11. X. Z. Ding, Z. Z. Qi and Y. Z. He, "Effect of Hydrolysis Water on the Preparation of Nano Crystalline Titania Powder via a Sol-Gel Process," Journal of Materials Science Lettters, Vol. 14, 1995, pp. 21-22.

12. S. J. Kim, S. D. Park, Y. H. Jeong and S. Park, "Homogeneous Precipitation of TiO_2Ultrafine Powders from Aqueous $TiOCl_2$ Solution," Journal of American Ceramic Society, Vol. 82, No. 4, 1999, pp. 927-932.

13. Q. Chen, Y. Qian, Z. Chen, G. Zhou and Y. Zhang, "Preparation of TiO_2 Powders with Different Morphologies by an Oxidation Hydrothermal Combination Method," Materials Letters, Vol. 22, 1995, p. 77-80.

14. Y. F. Chen, C. Y. Lee, M. Y. Yeng and H. T. Chiu, "The Effect of Calcination Temperature on the Crystallinity of TiO_2 Nanopowder," Journal of Crystal Growth, Vol. 247, 2003, pp. 363-370.

15. M. N. Khan, K. Shahzad and J. Bashir, "Thermal Atomic Displacements in Nanocrystalline TItanium Dioxide Studied by Synchrotron X-Ray Diffraction," Journal of Physics D: Applied Physics, Vol. 41, No. 8, 2008, Article ID: 085409.

16. U. K. Srdic, V. Wiedenman, M. Winterer and H. Hahn, "Nono Sized Ceramics of Coated Alumina and Zirconia Analysed with SANS," Journal of Applied Crystallography, Vol. 33, 2000, pp. 483-487.

17. J. Kohlbrecher, "SASFITTCL (2) Version 0.2 Alpha," PSI, Zurich, 1999. [Paper reference:1]

18. E. J. Shin, B. S. Seong, Y. S. Han, K. P. Hong, C. H. Lee and H. J. Kang, "Effect of Precipitate Size and Dispersion on Crystallization Behaviour in Ti-Added Ultra Low Carbon Steel," Journal of Applied Crystallography, Vol. 36, No. 3, 2003, pp. 624-628. J. Blasing, P. Kohlert, M. Zacharias and P. Veit, "X-Ray Fine Structure Investigations of Germanium Nanoclusters," Journal of Applied Crystallography, Vol. 31, 1998, pp. 589-593.

19. S.-J. Kim, J.-K. Lee, H.-G. Lee, S.-J. Kim and K.-S. Lee, "Photocatalytic Properties of Rutile TiO_2 Acicular Particles in Aqueous 4-Chlorophenol Solution," Journal of Materials Research, Vol. 18, No. 3, 2003, pp. 729-732.

7

PROBING GIANT MAGNETISM IN FE-N THIN FILM BY POLARIZED NEUTRON REFLECTOMETRY

Xiaowei Zhang[1, 2], Nian Ji[1, 2], Valeria Lauter[3], Haile Ambaye[3] and Jian-Ping Wang[1, 2]

[1] Electrical and Computer Engineering Department and the Center for Micromagnetics and Information Technologies (MINT), University of Minnesota, Minneapolis, Minnesota, USA

[2] Physics Department, University of Minnesota, Minneapolis, Minnesota, USA

[3] Neutron Science Scattering Division, Oak Ridge National Laboratory, Oak Ridge, Tennessee, USA

ABSTRACT

With the development in hard drive and permanent magnet industry, higher saturation magnetization (Ms) or magnetic induction (Bs) material is on high demand. According to the Slater-Pauling curve, the highest Bs value is ~2.45 T, which belongs to FeCo alloy. However, in 1972, Kim and Takahashi [1] announced that the new material Fe-N thin film exhibited an increase in the Bs value of 18%. From then on, tons of research works [2, 3, 4] had been dedicated in this area with conclusions on both sides, achieving either high Bs or low Bs. Among those works, the Bs values were obtained from the measurement of the thin film magnetic moment and its volume, which might cause a considerable amount of error depending on the accuracy of the film thickness and area measurement. Other concerns also include the uncertainty of Ms value due to the subtraction of Fe underlayer. Indeed, a direct measurement of Bs is the key to clarify the discrepancies between these results. Here, we are presenting the method of polarized neutron reflectometry (PNR) to measure the Bs of the partially ordered Fe16N2 thin film. PNR allows the interface magnetism study, [5] the absolute magnetization determination and magnetic depth profile in single thin films, [6, 7, 8, 9] and complicated superlattice structures. [10, 11] In the following scenarios, different partially ordered Fe16N2 thin films are fabricated and are measured using PNR subsequently. Their Bs values are obtained using the fitting results of nuclear scattering length density (NSLD) and magnetic scattering length density (MSLD). Also, a PNR application on FeN thin film will also help us to understand the switching picture of the FeN thin film with external applied field.

Keywords: $Fe_{16}N_2$, high saturation magnetization, polarized neutron reflectivity, thin film

1. INTRODUCTION

1.1. Pnr Introduction

The interaction of neutrons with matter can be described in the optical formalism using the concept of reflective index for the medium. This can be obtained by Schrödinger equation, where the interaction of the neutron with medium is denoted by V:

$$\left[-\frac{\hbar^2}{2m}(\Delta + k_0^2) + V\right]\Psi = 0 \tag{1}$$

$$\frac{\hbar k_0^2}{2m} = \omega \tag{2}$$

where m is the neutron mass, $k0$ is the wave vector in vacuum, $\hbar\omega$ is its energy. So the refractive index can be written as:

$$n^2 = 1 - V / \hbar\omega \tag{3}$$

Here, we can define the strong interaction potential with born approximation, which can be described by the Fermi pseudopotential: [12]

$$V_F(r) = b\left(\frac{2\pi\hbar^2}{m}\right)\delta(r), \tag{4}$$

where b is the scattering length and r is neutron position. The mean potential of the interaction between neutron and medium is the given by the integration of the space:

$$V = \frac{1}{V}\int V(r)d^3r = \frac{2\pi\hbar^2}{m}Nb, \tag{5}$$

where N denotes the number density of the atoms. The multiplicity of number density and scattering length is called scattering length density (SLD) ρ.

2. SPECULAR REFLECTIVITY OF NEUTRONS FROM INTERFACE

If we compare the situation of neutron reflection to the light propagation at the interface of different mediums, the different interaction potential of neutron with medium can be seen as different mediums with different refraction indices. This follows Fresnel equations (Fig. 1).

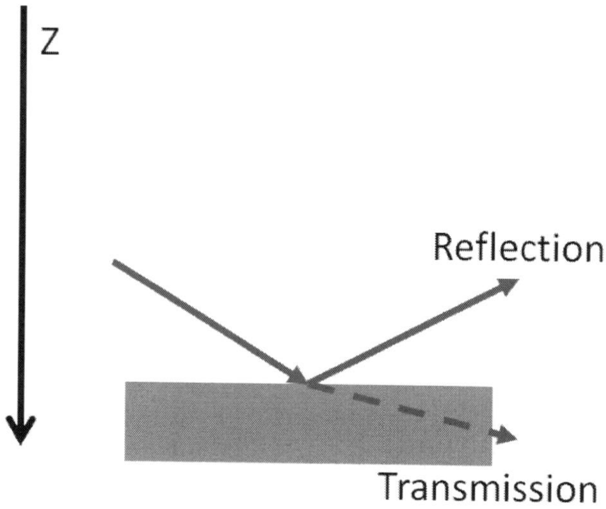

Figure 1. A schematic illustration of neutron interacting with matters at the interface obeys the Fresnel equations in analogy to the optical formalism.

$$R = |r|^2 = \left| \frac{k_1 - k_2}{k_1 + k_2} e^{i2k_1 z} \right|^2$$

$$T = |t|^2 = \left| \frac{2k_1}{k_1 + k_2} e^{i2(k_1 - k_2)z} \right|^2 \qquad (6)$$

For bulk materials, after applying Snell-Descartes' law, the reflection coefficient at a small incident angle can be written as a function of the wave-vector transfer q:

$$R(q) = \left| \frac{q_z - \sqrt{q_z^2 - q_c^2}}{q_z + \sqrt{q_z^2 - q_c^2}} \right|^2, \qquad (7)$$

where qc is the wave vector, which corresponds to critical angle of total external reflection. It is also easy to show that at high q limit, the reflectivity R follows the $1/q^4$ power law (Fresnel decay).

Dynamic calculation can also be applied with Born approximation. The reflectivity is equal to the Fourier transformation of the SLD profile: [13]

$$R = \frac{1}{q_z^2}\left|\int_{-\infty}^{\infty} e^{iq_z y}\rho(y)\,dy\right|^2 . \tag{8}$$

It is worth mentioning that Fresnel decay originates from a planar surface and the decay will result in a extremely low intensity at high q value, which will be very difficult to collect with the mixture of background noise or incoherent scattering from the substrate.

The critical angle θc is usually small and can be written as follows at the total reflection condition:

$$\cos\theta_c = n = 1 - \frac{\lambda^2}{2\pi} Nb, \tag{9}$$

where λ is the wavelength. It is useful to use Taylor expansion. Therefore, the critical angle θc and the corresponding critical wave vector qc are

$$\theta_c = \sqrt{\frac{Nb}{\pi}}\lambda \tag{10}$$

$$q_c = 4\sqrt{\pi Nb}. \tag{11}$$

Usually the thickness of the thin film is larger than the wavelength of neutron, and so the reflecting beams of the neutron will give constructive/destructive patterns according to Bragg's diffraction law.Figure 2 is an example of a reflectivity from a bare bulk substrate and of a typical thin film:

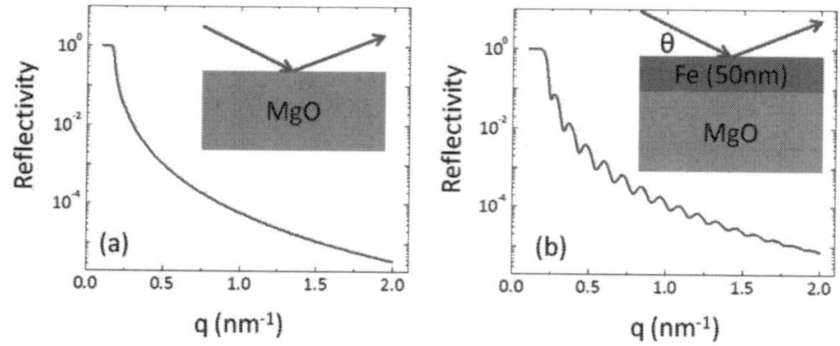

Figure 2. Simulated neutron reflectivity curves of (a) a flat MgO substrate and (b) a flat 50-nm-thick Fe thin film grown on MgO substrate.

3. NEUTRON REFLECTIVITY FROM A MAGNETIC INTERFACE

In the presence of an external magnetic field, BB, Zeeman energy will be present due to the spin of neutron itself:

$$V_Z = -g_n \mu_n B, \tag{12}$$

where gnμn gives the neutron magnetic moment. Since B=μ0H+MB=μ0H+M, where *M* is the magnetization of the material, we can express the potential in terms of *M*. At the magnetic interface with the addition of this energy, the total interaction potential energy between the neutron and medium can be written as follows:

$$V = \frac{2\pi\hbar^2}{m}(\rho_n \pm \rho_m), \tag{13}$$

where ρn is the nuclear SLD of the medium and ρm is the magnetic SLD of the medium, which can be defined by:

$$\rho_m = -\frac{m}{2\pi\hbar^2}\mu_n \sum M. \tag{14}$$

The sign ± represents spin-up and spin-down cases corresponding to the magnetization of the medium. In respect to neutron polarization, this shows that the magnetic SLD is directly related with the magnetization in the medium that the neutron is shining upon, which gives a direct measurement of the saturation magnetization when saturated in the external field.

Similarly, the critical angle θc and corresponding critical wave vector qc can be written as:

$$\theta_c = \sqrt{\frac{\rho_n \pm \rho_m}{\pi}}\lambda \tag{15}$$

$$q_c = 4\sqrt{\pi(\rho_n \pm \rho_m)}. \tag{16}$$

Therefore, spin-up reflectivity will have a large critical angle/momentum transfer compared to spin-down reflectivity (Fig. 3). If we present the potential in the matrix form with Pauli operator, then:

$$V = \frac{2\pi\hbar^2}{m}\begin{pmatrix} \rho_n & 0 \\ 0 & \rho_n \end{pmatrix} \mp \mu_n \begin{pmatrix} B_z & B_x - iB_y \\ B_x + iB_y & -B_z \end{pmatrix} = \frac{2\pi\hbar^2}{m}\begin{pmatrix} \rho_n + \rho_{mz} & \rho_{mx} - i\rho_{my} \\ \rho_{mx} - i\rho_{my} & \rho_n - \rho_{mz} \end{pmatrix}. \tag{17}$$

The potential we obtained above is the diagonal terms, which corresponds to the non-spin flip situation. The off diagonal terms [14]will represent the spin flip situation, which will not be discussed in this article.

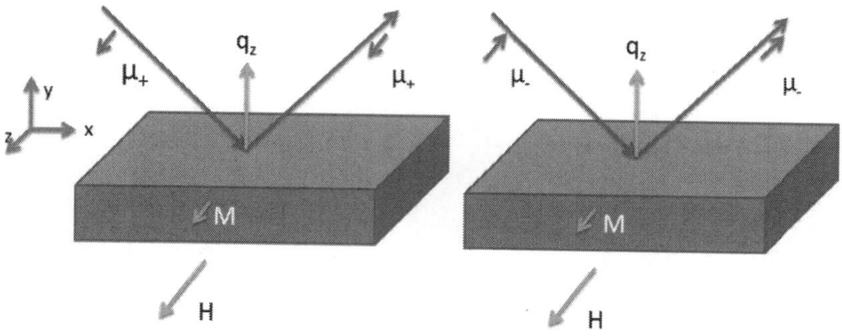

Figure 3. Polarized neutron reflectivity with two different spin polarizations. The spin-up (μ+) and spin-down (μ–) neutrons are indexed as parallel and anti-parallel with respect to the external field direction.

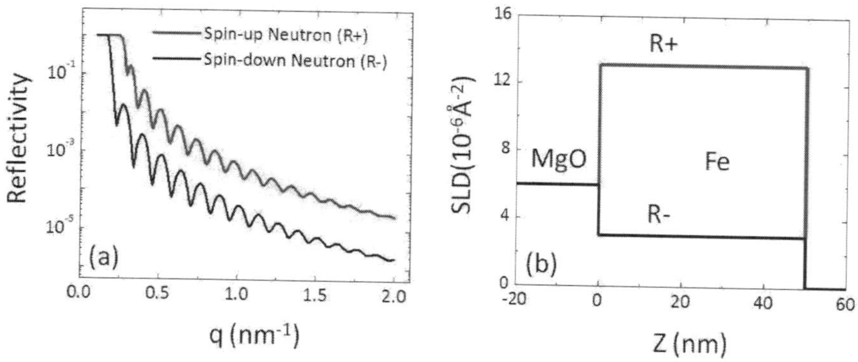

Figure 4. (a) The simulated reflectivities of a 50-nm Fe thin film grown on MgO substrate after saturation in-plane with external field. (b) The depth-dependent SLD according to the reflectivity curves in (a).

As seen from the reflectivity curves (Fig. 4), the magnetic interaction splits the non-magnetic oscillation curve into two branches. Fitting the reflectivity will yield the SLD for each spin case:

$$\rho_{tot+} = \rho_n + \rho_m$$
$$\rho_{tot-} = \rho_n - \rho_m. \quad (18)$$

We can solve for ρm and, therefore, retrieve the Ms of the sample.

4. GIANT BS INDUCED BY STRAIN EFFECT ON EPITAXIAL $FE_{16}N_2$ THIN FILM [15]

Two sets of partially ordered $Fe_{16}N_2$ thin film samples are fabricated on MgO substrate with facing target sputtering system. After the deposition of Fe underlayer at 300°C, a Fe-N layer is grown subsequently on the top of Fe layer at room temperature with the optimum volume ratio of N_2 and Ar mixture so that the stoichiometry in Fe-N layer is Fe/N=8:1. An in-situ annealing is carried out right after at 120°C for 20 hours in vacuum. To explore the in-plane tensile strain effect on the Bs value of Fe-N thin film, two samples are grown with different Fe underlayers, 2 nm and 20 nm, respectively.

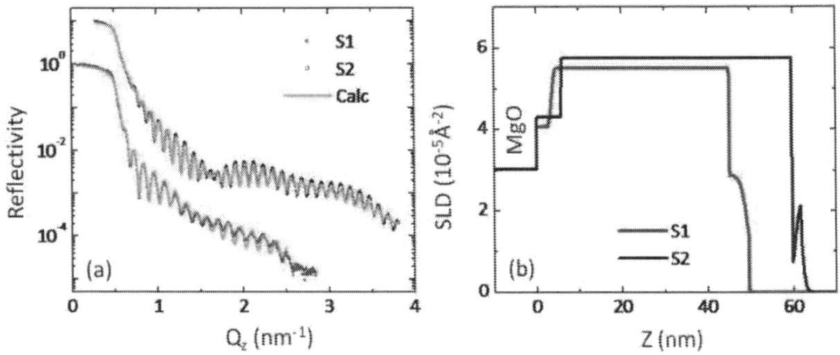

Figure 5. X-ray reflectivity characterization. (a) The fitted x-ray reflectivity curves measured on samples S1 and S2 (vertically offset by a factor of 10). (b) Calculated depth-dependent x-ray SLD profiles.

Both samples are investigated by PNR using the Magnetism Reflectometer at Spallation Neutron Source at Oak Ridge National Laboratory (SNS ORNL). [16] The reflectivities with the spin of the neutrons being either parallel (R^+) or anti-parallel (R^-) to the applied magnetic field are collected simultaneously. From these data the depth profiles of the SLD of both NSLD and MSLD are obtained. The PNR experiments are performed at room temperature in the saturation external field of H=1.0 T applied in-plane of the sample. The R^+ and R^- reflectivity data are fitted simultaneously using a genetic algorithm with an exact recursive matrix calculation embedded in the *Simulreflec 1.0* package. [17] The experimental reflectivity and calculated curves with best chi-squared fit for samples S1 and S2 are shown on Fig. 6a and b, respectively. Their corresponding structural NSLD and magnetization depth profiles are plotted in Fig. 6c and d.

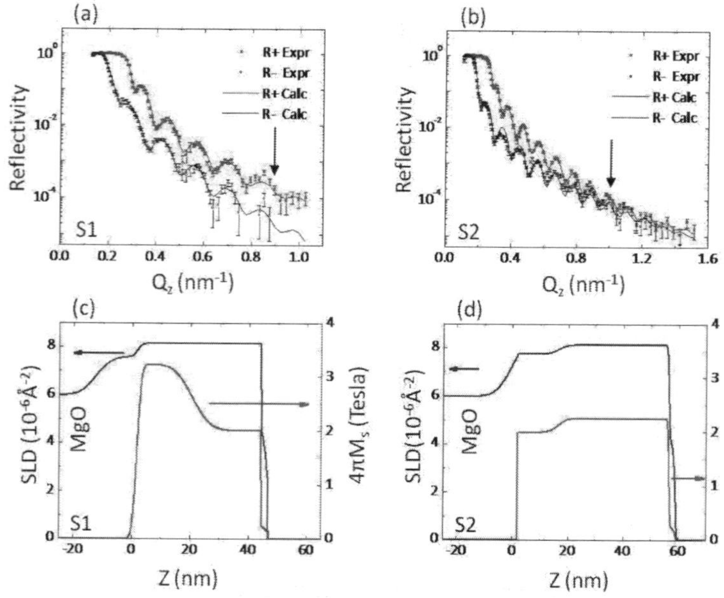

Figure 6. Polarized neutron reflectivity characterization. (a) and (b) Experimental polarized neutron reflectivities together with the fitted curves as functions of momentum transfer Q for samples S1 and S2 as labeled, respectively. The arrows at high Q region (>0.8nm^{-1}) indicate the difference of the magnetic properties toward the bottom interface between these two samples. (c) and (d) Structural (Brown) and magnetic (Green) depth profiles for samples S1 and S2 as labeled correspondingly.

In data modeling process, the structural NSLD and layer choice were constrained to closely match X-ray Reflectivity (XRR) results (Fig. 5). To account for the possibility that the film possess homogeneous chemical composition but potentially different magnetization, the Fe-N layer was subdivided into three slabs where NSLD are forced to be identical but thickness, roughness, and MSLD were allowed to vary independently. When fitting the data to extract the physical depth profile of the film structure, the only free parameter that was allowed to vary was the magnetization as other parameters were already framed by the XRR result and only marginal improvement was possible. After the comparison of the XRR and PNR result, we can find a relatively large roughness at high Q range, which is most likely due to either not enough data acquisition at this specific range or the chemical compound that is formed at the interface during the thin film growth or annealing process, such as MgO:N.

It is very obvious that an anomalously large magnetization is formed at the bottom of the film with a range of about 20 nm. The MSLD for this part of the film is 7.2~7.5 × 10^{-6}Å^{-2}, which in turn means the Ms value is 3.1~3.2 T. This value is significantly higher than bulk Fe (40~50%) and $Fe_{65}Co_{35}$(20~30%). When the SLD profile extends to higher layer of the film, the MSLD values drops to 4.66 × 10^{-6}Å^{-2} and represents a low Ms value of 2.01 T. This non-uniformity of the magnetization across the film cannot be replaced by single layer model, which will fail to resemble the reflectivity behavior of the experiment. For sample S2, the resulted MSLD is close to 5 × 10^{-6}Å^{-2} for the Fe-N layer, corresponding to Ms of 2.15 T, which does not show the presence of giant Bs. The different Bs results of the two samples are due to the strain that is experienced by the epitaxial growth of the Fe-N layer. In sample S1, larger in-plane tensile strain contributes to high Bs.

Further justification for the different magnetic structure upon Fe buffer thickness change comes from the spin asymmetry (SA) $(R^+-R^-)/(R^++R^-)$ plot shown in Fig. 7. SA of both samples were fitted with experiment data. Initially, the SA values of both samples are oscillating at a high value. However, the SA value of sample S2 drops rapidly close to zero, whereas that of sample S1 almost reaches unity. In this high scattering vector region (q>0.8nm^{-1}), the interface of the bottom layer will dominate MSLD and NSLD. To account for the disparity of the different behaviors of two samples, |MSLD–NSLD| >> MSLD needs to be satisfied for sample S1. Since NSLD is similar for both samples, this observed feature directly proves the substantial enhancement of MSLD in the high Ms sample comparing to that of S2 at the bottom interface.

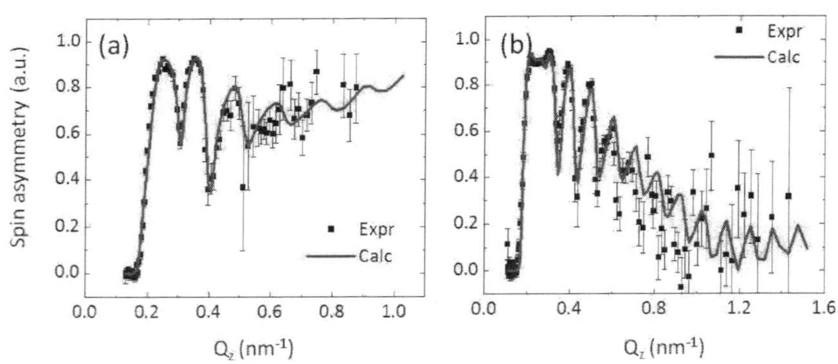

Figure 7. Spin asymmetry plotting of samples S1 (a) and S2 (b). It is clear as it approaches high q, the trend of the curves are different for these two samples.

5. HIGH BS OBTAINED ON PARTIALLY ORDERED $FE_{16}N_2$ MULTILAYER THIN FILM [16]

Multilayer structures are very common in many areas that they can bring many new physical phenomena. To explore the possibility of applying $Fe_{16}N_2$ in multilayer structure, two samples are fabricated on GaAs substrate with Fe/Fe-N layer stack. Samples L1 and L3 are with 1 and 3 repetitions, respectively.

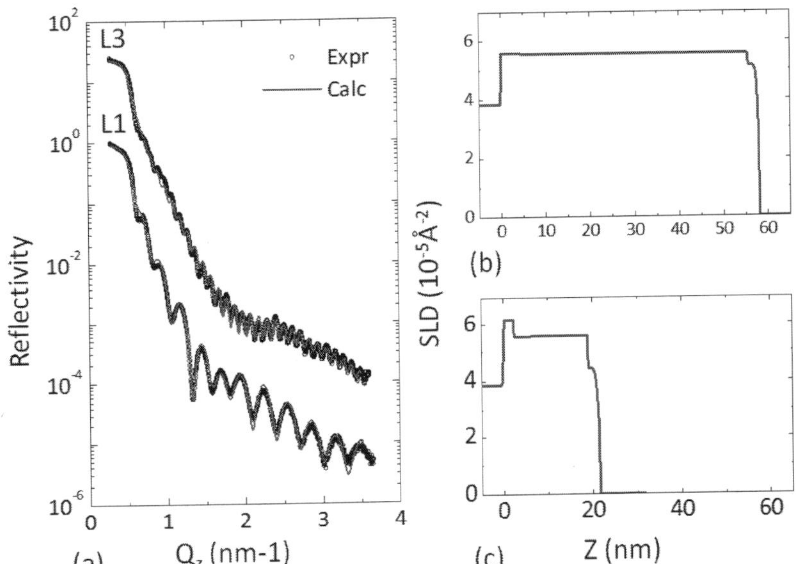

Figure 8. (a) X-ray reflectivity of $Fe_{16}N_2$ samples L1 and L3. (b) Electron SLD of L3 was homogeneous through the whole sample. (c) SLD of L1 stayed the same after the first ~2 nm bump.

It is known that co-refining the x-ray and neutron reflectivity curves allows an unambiguous determination of magnetic depth profile. [19, 20] When modeling the chemical part of the film structure, consistent structure for fitting both samples is set the same for both PNR and XRR. In this case, only marginal adjustment is allowed during the fitting process. The two layer model for L1 and four layer model for L3 work well in both fitting, which illustrates the neutron scattering of Fe and N elements that accounts for the difference for each layer. It is also found out that the N concentration at the interface region is lower than that before annealing process.

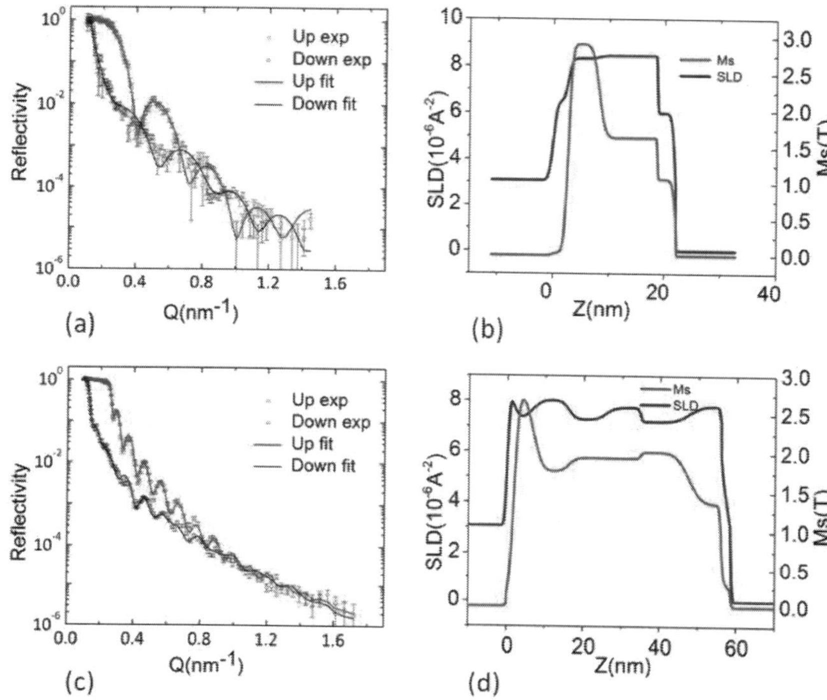

Figure 9. (a) Experimental and fitted reflectivity of sample L1. (b) Sample L1 SLD and Ms in-depth profile after fitting. (c) Experimental and fitted reflectivity of sample L3. (d) Sample L3 SLD and Ms in-depth profile after fitting. Both (b) and (d) curves exhibit high Ms value at bottom part of the film for ~5 nm scale.

PNR was conducted on these two samples, respectively, with field applied in-plane to saturate the magnetic moment. It is important to notice that these samples are different from the ones with optimum FeN atomic ratio (Fe/N=8/1), whose Ms values are homogeneous throughout the film according to PNR measurement. In Fig. 9, reflectivity curves showed different oscillatory periods due to the differences in film thickness. Green lines of both samples in Fig. 9 (b) and (d) indicate the giant Ms at the bottom of the film, with a thickness of 5 nm, respectively. Since all the other FeN phases possess low Ms values, the only option for this enhancement should come from $Fe_{16}N_2$. Considering the fact that Fe seed layer is deposited before FeN layer, N atoms diffusion is most likely to occur to form $Fe_{16}N_2$ at the bottom as N atoms are initially rich on the top of Fe underlayer. $Fe_{16}N_2$ is formed at this specific location near the substrate with the right stoichiometric amount of N atoms. On the top of this giant Ms layer, a N rich phase is expected since N atoms are initially rich and a lower Ms result is obtained by PNR. This transition of Ms from high to low also occurs on L3. A reduced but still "giant" magnetization resides at the bottom interface. Compared to sample L1, the transition from high Ms to low Ms is comparably slow. Considering the XRD analysis achieved above, α"-$Fe_{16}N_2$ and α'-Fe-N martensites are present in the film structure. Since high Ms behavior is obtained, these observation on PNR analysis can only be explained by the presence of high Ms $Fe_{16}N_2$ phase and N rich phase. The formation of these mixture phases is most likely due to the strain that the film experiences with thinner Fe underlayer on the substrate. Multilayer structure of Fe/FeN provides the same environment of single Fe/FeN layer, but the difference is that higher Fe/FeN layers, which are not next to substrate, do not exhibit giant Ms. This is a result of the lattice strain relaxation in the film norm direction. To better demonstrate the fitting of PNR, chemical structure extracted from NSLD was checked and the corresponding model also fitted XRR data. It is noticeable that the bottom part of samples L1 and L3 have different electron and NSLD from XRR and PNR result. This region is most likely Fe and GaAs compound due to the inter-diffusion of Fe and GaAs atoms at their interface. The addition of this layer actually improved our fitting to the experiment data and was confirmed by SLD depth profile.

6. USING PNR TO VISUALIZE THE HIGH PERPENDICULAR CRYSTALLINE ANISOTROPY [19]

The ordered $Fe_{16}N_2$ unit cell possesses tetragonality that gives rise to a high crystalline anisotropy, which is perpendicular to the film plane. This can be measured using vibrating sample magnetometer (VSM) in Fig. 10. However, VSM will not give detail information that will illustrate the FeN layer behavior in presence of the applied magnetic field.

During the PNR experiment, the external field is applied in the film plane direction. Both spin configuration signals are collected simultaneously. Fig. 11a–c show the reflectivity curves with fitting for spin-up and spin-down cases in three different external field intensities. At an external field of 20 kOe, the whole film magnetic moment will be saturated in the external field direction.

Since the same sample is tested in different fields, the chemical part, namely NSLD, stays the same during the fitting process. Three different MSLD depth profiles are generated after fitting in Fig. 11e. Since only one field intensity is strong enough to saturate the film in the film plane direction, the MSLD represent the in-plane component of the magnetization. It is clearly seen that the MSLD value drops rapidly for Fe-N layer, compared to Fe layer, when the external field is lowered. Since Fe layer is magnetically soft, all three fields will keep the Fe moment aligned in-plane. For Fe-N layer, there could be two scenarios to consider. First, the Fe-N film consists of antiferromagnetic domains in the film plane, which means the Fe-N layer has uniaxial in-plane anisotropy. This would directly contradict the fact that in-plane M-H loop does not depend on the in-plane orientation. Also, in this case, a strong off-specular signal should be observed, which is not our case either. Second, the magnetic moment turns out-of-plane due to perpendicular easy axis. This does explain our result since we do not observe off specular signal. When external field is lowered, the out-of-plane component of Fe-N film parallels the neutron beam momentum transfer and in-plane component contributes to the specular reflectivity. Therefore, the reduction of the magnetization in the in-plane direction after lowering external field can be explained by the out-of-plane easy axis of the Fe-N film.

Also, it is worth to mention that the oscillation amplitude of the reflectivity curves for R^- curve is much smaller than that of R^+ curve. This is most likely due to the modulation of the external field that the SLD is getting close to that of the substrate. This phenomenon is more robust when it approaches the Fe/GaAs interface where the reflectivity behavior is most dominated by this region at high q_z.

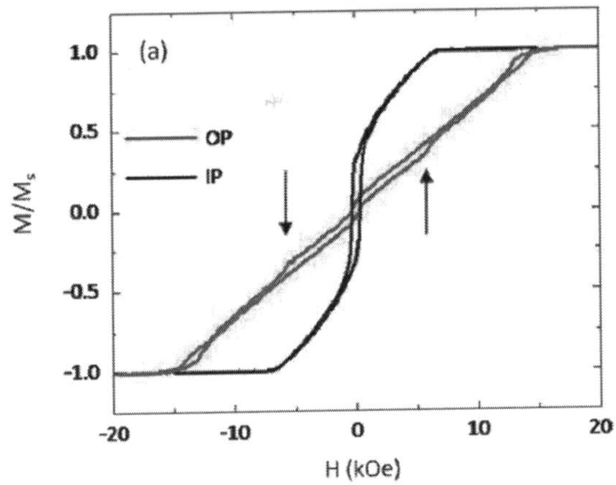

Figure 10. In-plane (black) and out-of-plane (red) M-H loops measured on one partially ordered $Fe_{16}N_2$sample. The Fe-N layer switches at Hc~5.7 kOe (black arrows).

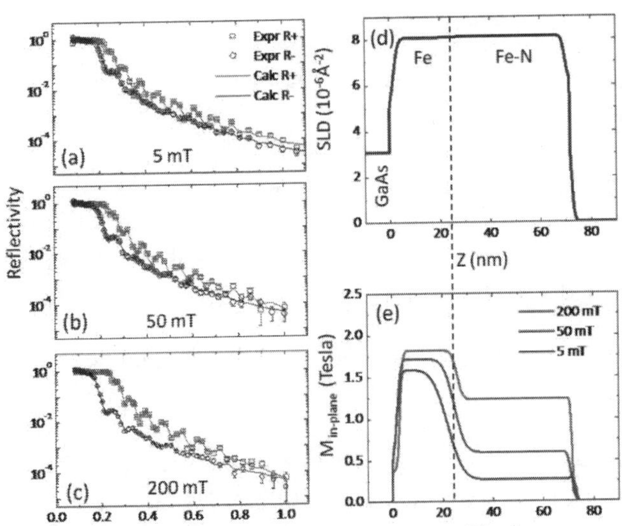

Figure 11. (a)–(c) Fitted reflectivity curves with experiment data of Fe-N sample at external field of 5, 50, and 200 mT, respectively. (d) NSLD depth profile of Fe-N sample and (e) MSLD, the in-plane magnetization component, depth profile of Fe-N sample at above three different external fields.

7. SUMMARY

By using PNR, the Bs value can be accurately obtained directly in contrast to the conventional way that involves the thin film volume. In addition, we are also obtaining the Fe-N thin film NSLD and MSLD in-depth profile, which helps us to understand the physics behind the high Bs value. The above results are all derived from the specular reflections of the polarized neutron beam. This will facilitate the understanding of the giant Bs partially ordered $Fe_{16}N_2$ thin film in the film plane direction.

8. ACKNOWLEDGEMENTS

This work was partially supported by ARPA-E (Advanced Research Projects Agency- Energy) project under Contract No. 0472-1595, Seagate Technology and Western Digital. Parts of this work were carried out in the Characterization Facility through NSF MRSEC program at University of Minnesota.

REFERENCES

1. T. K. Kim and Takahashi, Appl. Phys. Lett. 20, 492 (1972).
2. Y. Sugita, H. Takahashi, M. Komuro, K. Mitsuoka and A. Sakuma, J. Appl. Phys. 76, 6637 (1994).
3. M. A. Russac, C. V. Jahnes, E. Klokholm, J. Lee, M. E. Re, and B. C. Webb, J. Appl. Phys. 70, 6427 (1991).
4. J.-P. Wang, N. Ji, X. Liu, Y. Xu, C. Sánchez-Hanke, Y. Wu, F. M. F. de Groot, L. F. Allard , and E. Lara-Curzio Depa.
5. J. Chakhalian, J. W. Freeland, G. Srajer, J. Strempfer, G. Khaliullin, J. C. Cezar, T. Charlton, et al., Nat. Phys. 2, 4 (2006) 244–248.
6. J. A. C. Bland, D. Pescia, and R. F. Willis, Phys. Rev. Lett. 58, 1244 (1987).
7. M. A. M. Gijs, F. Petroff, Magnetic Ultra Thin Films, Multilayers and Surfaces Elsevier, 1997.

8. J. A. C. Bland, J. Lee, S. Hope, G. Lauhoff, J. Penfold, D. Bucknall, J. Magn. Magn. Mater. 165, 46 (1997).

9. S. J. Blundell, M. Gester, J. A. C. Bland, H. J. Lauter, V. V. Pasyuk, and A. V. Petrenko, Phys. Rev. B 51, 9395 (1995)

10. J. A. Borchers, J. A. Dura, J. Unguris, D. Tulchinsky, M. H. Kelley, C. F. Majkrzak, S. Y. Hsu, R. Loloee, W. P. Pratt Jr., and J. Bass, Phys. Rev. Lett. 82, 2796 (1998).

11. J. Hoppler, J. Stahn, C. Niedermayer, V. K. Malik, H. Bouyanfif, A. J. Drew, M. Rössle, A. Buzdin, G. Cristiani, H. U. Habermeier, and B. Keimer, Nat. Mater. 8, 315 (2009).

12. X. L. Zhou and S. H. Chen, Phys. Rep. 257, 223–348 (1995).

13. M. R. Fitzsimmons and C. F. Majkrzak, Application of Polarized Neutron Reflectometry to Studies of Artificially Structured Magnetic Materials. pp 15

14. N. Ji, V. Lauter, X. Zhang, H. Ambaye, and J. P. Wang. Appl. Phys. Lett. 102, 072411 (2013).

15. V. Lauter-Pasyuk, H.J. Lauter, B. Toperverg, O. Nikonov, E. Kravtsov, L. Romashev, V. Ustinov, J. Magn. Magn. Mater., 226 Part 2, 1694 (2001).

16. V. Lauter, et. al., Physica B 404, 2543 (2009).

17. http://wwwllb.cea.fr/prism/programs/simulreflec/simulreflec.html.

18. X. Zhang, N. Ji, V. Lauter, H. Ambaye, and J. P. Wang, J. Appl. Phys. 113, 17E149 (2013).

19. G. P. Felcher, Phys. Rev. B 24, 1595 (1981).

20. C. F. Majkrzak, Physica B 173, 75 (1991).

21. N. Ji, M. S. Osofsky, V. Lauter, L. F. Allard, X. Li, K. L. Jensen, H. Ambaye, E. Lara-Curzio and J.-P. Wang, Phys Rev B 84 (24), 245310 (2011).

8

STRUCTURAL IRREVERSIBILITY AND ENHANCED BRITTLENESS UNDER FATIGUE IN ZR-BASED AMORPHOUS SOLIDS

Peng Tong [1,2], Despina Louca [1]

[1]Department of Physics, University of Virginia, VA 22904, USA

[2]Key laboratory of Materials Physics, Institute of Solid State Physics, Chinese Academy of Sciences, Hefei 230031, China

ABSTRACT

The effect of fatigue on ZrCuAl amorphous metals induced by mechanical cyclic loading is investigated using inelastic neutron scattering and the pair density function analysis of neutron diffraction data. With cooling, the local atomic structure undergoes reorganization under fatigue that is directly related to the number of fatigue cycles. Also under fatigue, suppression in the atomic dynamics is observed as well. A structural restructuring occurs within a 4 Å radius and intensifies with increasing the compression cycles, whereas the vibrational density of states is attenuated as the intensity shifts towards the elastic, zero-energy transfer peak. The combined static and dynamic structural effects are a signature of the microscopic changes brought about by fatigue, and together may be the onset for subsequent behaviors following extended cyclic loading such as fracture. Even after the load is removed, the structural changes described here remain and increase with repeated cyclic loading which is an indication that the lattice deforms even before shear bands are formed.

Keywords: local structure; compression fatigue; dynamics; VDOS attenuation

1. INTRODUCTION

Amorphous metals are solids devoid of periodicity, prepared by the very fast cooling of a viscous liquid to avoid crystallization. Trapped in this supercooled state with an atomic topology that resembles that of the liquid [1], amorphous alloys exhibit uniquely different metallurgy from crystals, and have widespread industrial implications. With strengths that can surpass those found in crystalline counterparts, how they mechanically respond and deform under applied stresses is central to their functionality. Typically, metallic glasses have undesirably low ductility, though, and most commonly plastically deform through shear bands [2,3,4,5]. The structure, which is usually perceived as short-ranged in nature, and how it deforms under stress is clearly important in controlling plastic flow [6]. However its role and that of its dynamics are still poorly understood.

In spite of their high strengths [7] glassy metals are often brittle [8] like ceramics, instead of crystalline metals that are ductile, and permanently deform by forming shear bands under stresses. In the present work, we investigate the relation between the vibrational properties in the glassy state and the atomic properties leading to plastic flows [9] under mechanical testing such as compression fatigue. The system used for the purpose of addressing this question is a strong glass former, ZrCuAl [10,11,12]. The impact from fatigue on the nature of static disorder and low temperature dynamics are investigated via neutron scattering.

Amorphous matter exhibits unusual low temperature characteristics [13] that are distinctly different from any crystalline equivalent. One anomaly is the observation of excess vibrational density of states manifested in the so-called boson peak, and observed by techniques such as neutron [14] and Raman scattering [15]. This is also manifested in an unusual low temperature behavior of the heat capacity, C_P, as well, in which $C_P$deviates from the T^3 Debye law expected of a typical metallic solid with cooling. The boson peak is a universal attribute in glassy systems. Its origin has been linked to transverse vibrational modes associated with defective structures [16], to local vibrational modes of clusters [17], and to crossover dynamics of local and acoustic modes [18]. More recently it was proposed that the boson peak is analogous to the acoustic van Hove singularity in crystals [19,20,21]. The excess vibrational density of states, as determined both from the neutron inelastic scattering and C_P measurements displays a maximum of a few milli-electron Volts (meV) (of the order of 10^{-12} s). The intensity of the boson peak has in turn been directly related to other concepts such as fragility [1,22] and to the local free volume [16], which has been well characterized in molecular glasses [23]. However, little is known regarding the nature of the boson peak in metallic glasses with isotropic structures and even less so under the influence of mechanical deformation. With only a few of the alloys discovered thus far exhibiting good mechanical properties [24,25], it would be important to investigate the relation of static disorder and atomic dynamics to the mechanical deformation brought upon by fatigue.

Fatigue is critical in applications such as springs, in some sensors, actuators, and microelectrochemical systems that require lifetimes of 10^{12} cycles if not more [26]. However, fatigue damage can occur in metallic glasses at very low stresses, sometimes as low as 10% of the yield stress [27], limiting their use in structural applications [26,27]. Understanding the relation of the atomic structure to the mechanical deformation behavior brought upon by a static load is a challenging issue. That is because bulk metallic glasses can deform both homogeneously and inhomogeneously. At low temperatures, the deformation is inhomogeneous where plastic flow concentrates on shear bands, accompanied by a decrease of the viscosity and an increase of the free volume leading to strain softening [28]. Shear bands are observed under fatigue conditions due to fatigue softening [29]. It has been suggested that the formation of shear bands and fatigue softening are linked to the fracture of bulk metallic glasses during the fatigue process [30].

In this paper, we show that under high-cycle fatigue, the ternary ZrCuAl glass changes in part due to an atomic restructuring that suppresses fluctuations. A local reorganization of the atom packing is observed, where the changes intensify as the number of compression cycles increases. This is best observed by cooling the system down to low enough temperatures that reduces the smearing induced by thermal vibrations. Associated with these changes in the static structure is a softening of the lattice dynamics. The attenuation implies an increase in the sound velocity, suggesting that the glass becomes stiffer and more likely brittle under fatigue.

2. RESULTS AND DISCUSSION

2.1. Specific Heat

The Zr based ternary glass is a model system with exceptional glass forming ability. The low-temperatureC_P anomaly associated with the presence of an excess vibrational low frequency contribution is shown inFigure 1a for $Zr_{60}Cu_{30}Al_{10}$ and $Zr_{50}Cu_{40}Al_{10}$. Plotted in the figure is C_P with the electronic contribution, γT, subtracted and then normalized by T^3 [27]. Such normalization should have rendered a temperature independent curve for a solid that follows the Debye model of C_P, but instead a bump is observed that reaches a peak around 10 K. This behavior has been previously observed in metallic and other types of glassy systems and is associated with strong glasses [27,31]. While the bump is present in both compositions, it is distinctly more intense in $Zr_{60}Cu_{30}Al_{10}$ that in turn may indicate that the amplitude of excess vibrational density of states is higher in this alloy. To reproduce the bump, the C_P is typically fit by two terms, one originating from the Debye model of the C_P and an Einstein term of a single oscillator with an Einstein temperature, $\theta_E \sim 56$ K for $Zr_{60}Cu_{30}Al_{10}$ ($\theta_E \sim 63$ K for $Zr_{50}Cu_{40}Al_{10}$). By combining both terms, the resulting curve reasonably reproduces the bump at low temperatures as seen in Figure 1a. An Einstein oscillator represents a localized vibrational mode, and although rare in closed-packed structures, is ubiquitous in compounds with large voids or cage structures [32,33].

It has been suggested that atoms occupying these voids in metallic glasses are loosely bound and are thus free to rattle almost independently, giving rise to single oscillator modes. Also shown in Figure 1b is the C_P for the same sample after fatigue. Very small differences can be discerned between the dashed (post-fatigue) and solid (pre-fatigue) lines. TheC_P temperature dependence of the $Zr_{50}Cu_{40}Al_{10}$ alloy shows a very small enhancement but no visible differences can be seen in $Zr_{60}Cu_{30}Al_{10}$.

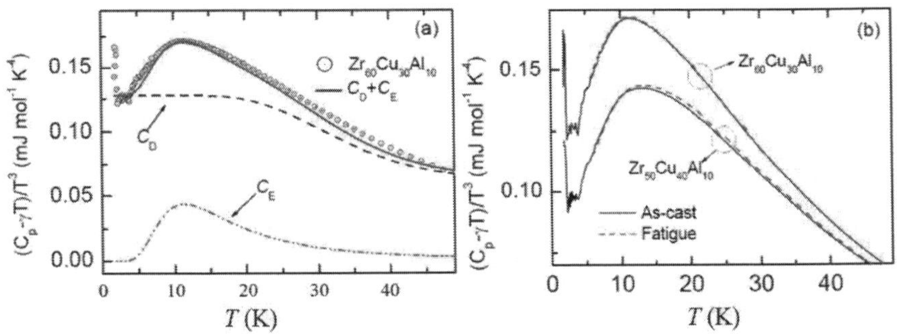

Figure 1. (**a**) The heat capacity at constant pressure, C_P, with γT subtracted and normalized by T^3 is plotted as a function of temperature for the two compositions. γ is the Sommerfeld constant; (**b**) The Debye,C_D and Einstein, C_E, terms are estimated. They are compared to the experimental C_P.

2.2. Inelastic Neutron Scattering

Shown in Figure 2a is the $S(Q)$ from data collected at 300 K for $Zr_{60}Cu_{30}Al_{10}$ and obtained from a diffraction experiment (solid line). The broad peaks are a signature of the static disorder present in the glass. This $S(Q)$ is used in the PDF analysis that will be discussed further below. Also shown in the same figure is the structure function obtained from the inelastic experiment but with the data integrated over the elastic peak from −2 meV to 2 meV (red symbols). The structure function obtained from the diffraction experiment extends to very high momentum transfer, Q, while the one obtained from the inelastic experiment is limited in Q, but the main features are reproduced as in the diffraction measurement.

Shown in Figure 2b is the energy, E–Q map collected at $T = 300$ K for $Zr_{60}Cu_{30}Al_{10}$ at the DCS. The red bar corresponds to the elastic scattering that has the brightest intensity. The intensity integrated in energy from −2 to 2 meV, the breadth of the elastic peak, yields the elastic $S(Q)$ shown in Figure 2a. Beyond the elastic peak, an inelastic continuum is observed below 2 Å^{-1} in addition to two strong localized modes. The experimental setup is such that most of the information we obtain will be in the neutron energy gain side (−meV) and this is what is shown in the figure. The neutron energy loss side (+meV) extends up about 10 meV. However, the elastic line is asymmetric, with the asymmetry appearing on the right side of the peak and most like due to the sample alignment. Because of this, the analysis was focused on the neutron gain side. Of the two modes, the rather narrow one is centered at Q~2.5 Å^{-1}and E_1~5 meV and appears at the same median Q-position as the strongest peak in the static $S(Q)$. Thus clearly the correlations associated with the tallest peak in the $S(Q)$ have the corresponding fluctuations in energy. Similarly, the broader excitation extends beyond Q~4 Å^{-1} with an E_2~15 meV and appears in the same Q-region as for the second broad peak in the $S(Q)$ of Figure 2a. By integrating the data over all accessible momentum transfers, the dynamic structure function, $S(\hbar\omega)$, shown in Figure 2c is obtained. The $S(\hbar\omega)$ is normalized by the Bose factor appropriate for the neutron energy gain side, $[n(\omega) + 1]$. The integration overall all Q incorporates all excitations seen in Figure 2b and shown in this figure is their temperature dependence. It has been previously suggested that such collective excitations result from acoustic, phonon-like modes arising from a pseudo-Brillouin zone due to some dynamic periodicity [34,35]. Below 50 K, the peaks are not discernible because of the low population occupancy, while their intensity increases following Bose-Einstein statistics. At 50 K, the E_1 peak at ~5 K is quite pronounced while E_2~15 meV is suppressed. Above 50 K, E_1 gains in intensity and by 300 K, it appears to be almost as intense as E_1. Shown in Figure 2d is the integration of the data in three different Q-slices corresponding to the three different regions in the E–Q contour map: from 0 to 2 Å^{-1} (low-Q), from 2 to 4 Å^{-1} (intermediate-Q) and from 4 to 7 Å^{-1} (high-Q) for the two compositions studied. In all the curves, the intensity for $Zr_{50}Cu_{40}Al_{10}$ is slightly lower than that for $Zr_{60}Cu_{30}Al_{10}$. Typically the phonon energy spectrum is normalized by the √M (the total mass) and hence as expected, the spectrum for the $Zr_{60}Cu_{30}Al_{10}$ is shifted closer to the elastic line. With temperature, E_1 remains essentially constant for both compositions, while E_2 shifts to lower energies by cooling from 150 K, indicating some kind of softening associated with these fluctuations.

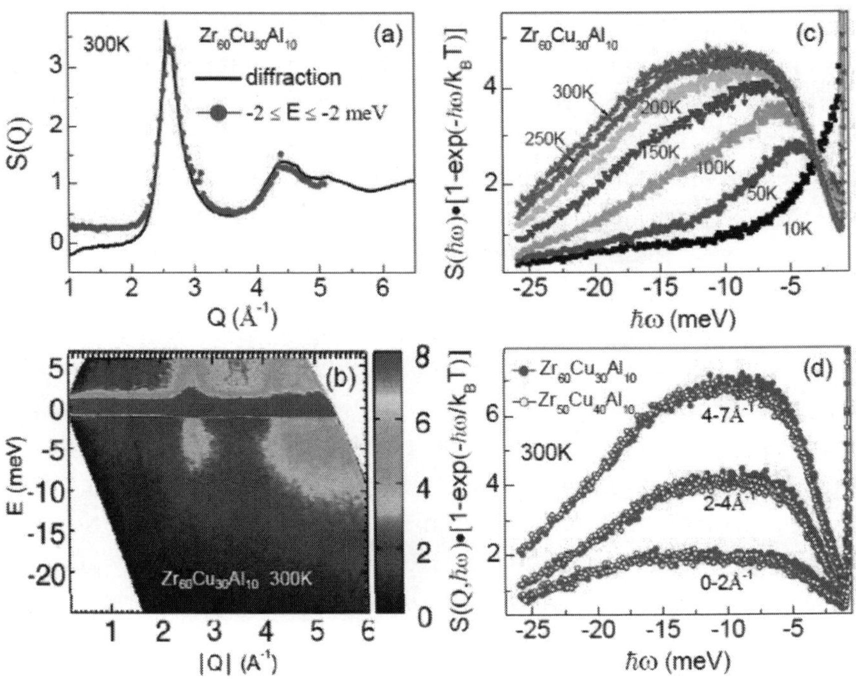

Figure 2. (**a**) The static $S(Q)$ determined from diffraction compares well to the elastic $S(Q)$ determined from the inelastic measurement. The small peaks observed in the elastic $S(Q)$ are from Al; (**b**) The energy*versus* momentum transfer map shows the presence of the two excitations. A broad continuum is also observed at very low Q; (**c**) The total $S(\hbar\omega)$ as a function of temperature for $Zr_{60}Cu_{30}Al_{10}$; (**d**)The $S(Q,\omega)$ determined at three regions of momentum space, low-Q, intermediate-Q and high-Q and compared between the two compositions studied.

2.3. Vibrational Density of States under Fatigue

What happens to the vibrational density of states (VDOS) intensity under fatigue? Shown in Figure 3a is the VDOS/E^2 for the two $Zr_{50}Cu_{40}Al_{10}$ alloys, one as-cast and the after fatigue. The VDOS function is given by $Z(\omega) = ((S(Q,\omega)\omega)/(Q^2[n(\omega) + 1]))$. An applied maximum stress of 1600 MPa was used at 10^5 compression cycles. Data were collected at two temperatures, 300 and 150 K. The amount of sample used for each experiment was different (50 g of the as-cast sample *versus* 10 g of the fatigue sample). Hence the data for each set of samples were normalized by the integrated intensity of the elastic peak and the high energy spectrum end was used to normalize the background. At constant temperature, the inelastic intensity is suppressed after fatigue while the intensity shifts towards the elastic line. The as-cast alloy consistently shows a higher intensity at both temperatures than the alloy after mechanical testing. Fatigue decreases the number of vibrational modes contributing to these excitations, in an analogous way to freezing of these modes as they become static. Shown in Figure 3b is the VDOS/E^2 at 300 K only and selectively integrated in the three Q regions defined earlier. In the low-Q and intermediate-Q regions, the intensity is clearly reduced after fatigue, and shifts to the elastic peak. However, in the high-Q region from 2 to 4 Å^{-1}, the intensity appears to change little with fatigue. This trend may indicate that the fluctuations mostly affected by fatigue involve long wavelength cluster correlations pertaining to shear bands.

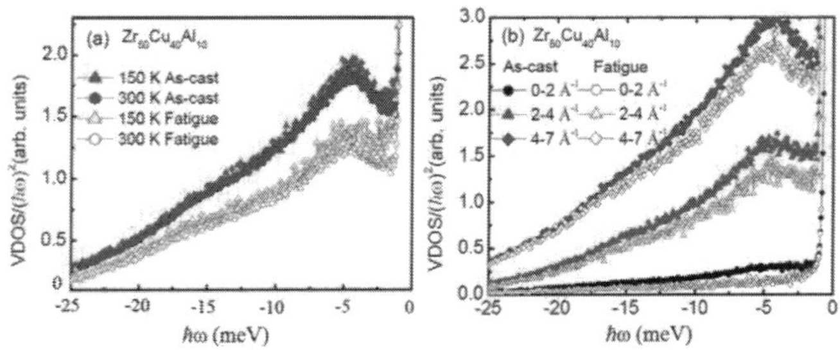

Figure 3. (**a**) The VDOS/E^2 for $Zr_{50}Cu_{40}Al_{10}$is shown for the data collected for an as-cast alloy and one that has undergone 10^5compression cycles. The inelastic intensity is suppressed at both temperatures shown. (**b**) The $S(Q,\omega)$ is determined at three regions of momentum space, low-Q, intermediate-Q and high-Q and the comparison is for data collected before and after fatigue.

2.4. Local Structure under Fatigue

The suppression of these excitations, similarly to the suppression of the boson peak [36] by external stimuli, may be linked to the structural re-organization that we observed in the local structure. Following a well established procedure of Fourier transforming the static $S(Q)$ from the reciprocal space into a real space representation, the pair density function (PDF) [37] is obtained and the results are shown in Figure 4. The PDF involves the integration over a finite energy window (~10 meV) and represents a time-averaged local structure, in the absence of an underlying periodicity. The neutron diffraction experiments were performed from 300 down to 10 K on as-cast alloys and under varying compression-compression fatigue cycles on the two compositions. The PDF up to 6 Å as a function of fatigue cycles is shown in Figure 4a for $Zr_{60}Cu_{30}Al_{10}$at a constant temperature of 10 K. With increasing the number of cycles, the height of the first peak rises while its width becomes narrow and its shape changes. A similar observation is made by cooling the structure from 300 to 10 K of the alloy subjected to 10^5 cycles (Figure 4b). Given that the area under the PDF peaks is conserved, it is not unusual for the peaks to become narrower and taller with cooling because cooling reduces the atomic thermal vibrations that contribute to the peak broadening. However, it is unusual for the peaks to change shape, as seen in both Figure 4a,b. It is even more unusual for the intensity to change at a constant temperature as seen in the PDF's of Figure 4a as a function of fatigue. Also shown in this figure are three partial functions, corresponding to the Zr-Zr, Zr-Cu and Cu-Cu atom correlations that contribute the most to the total PDF, and calculated from an embedded atom potential model [38,39]. The increase in the peak sharpness corresponds to the structure becoming more compact as the peak width narrows, and signals the reduction of free volume that may in turn be linked to brittleness.

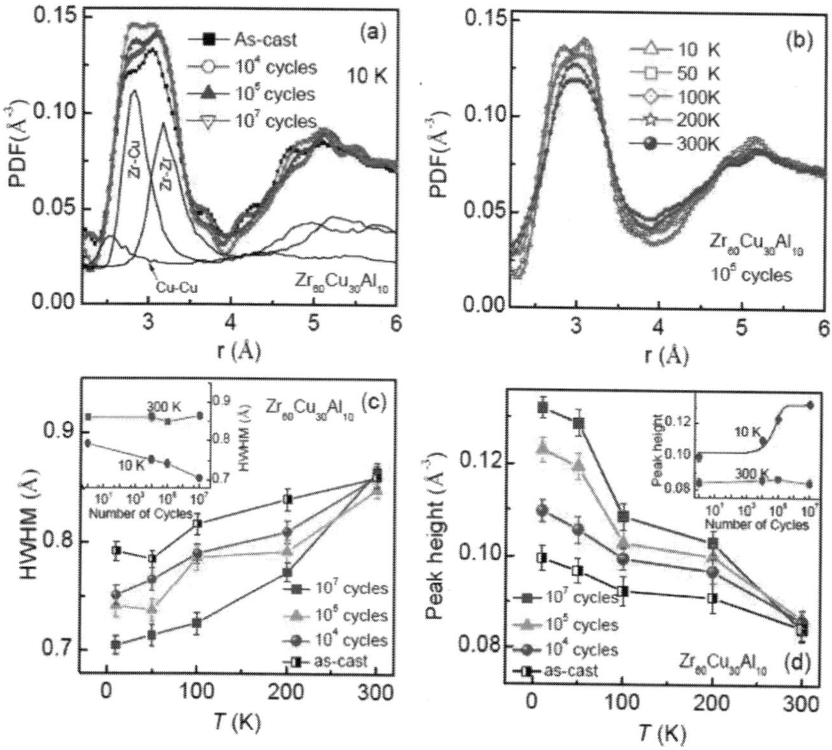

Figure 4. (**a**) The PDF corresponding to the local atomic structure of $Zr_{60}Cu_{30}Al_{10}$under different fatigue cycles at 10 K. Also shown are partial PDF's for select pairs obtained from molecular dynamics.(**b**) The local atomic structure as a function of temperature for the alloy at 10^5cycles.(**c**) The HWHM of the first PDF peak as a function of temperature for 4 different fatigue cycles. The inset shows the HWHM at two temperatures.(**d**) The change of the first PDF peak height with temperature at four fatigue cycles. The inset shows the change of the peak height at two temperatures.

The reduction of the half-width-half-max (HWHM) of the first peak, shown in Figure 4c, indicates that the first coordination sphere contracts upon fatigue loading. The changes are most noticeable at low temperatures than at room temperature (see inset of this figure) because of the reduction of thermal vibration, while a clear correlation is observed between the reduction of HWHM and fatigue cycles. A hysteresis is observed on warming where the system returns to its original state around 100 K (not shown). If we consider the temperature dependence of the as-cast sample as the baseline (black symbols), it is clear that the effects from fatigue are above this baseline. The change in the peak height of the first PDF peak at this maximum 1600 MPa stress level is most dramatic as the load cycles increase as shown in Figure 4d. In the inset of this figure, the data points at 10 K can be fit by an exponential function although there are only 4 points. But in comparison to the constant behavior observed at 300 K, the rapid peak height change at 10 K is clear. Thus permanent changes are observed in the local atomic structure in response to the presence of residual stresses even after the load is removed. On the other hand, at 300 K, the response remains almost constant where the effects are masked by the thermal vibration of atoms.

3. EXPERIMENTAL SECTION

The alloys of $Zr_{50}Cu_{40}Al_{10}$ and $Zr_{60}Cu_{30}Al_{10}$ (in atomic percent) were prepared using a ladle-hearth type arc-melt tilt-casting technique. The ingots were cut into 6.0 mm × 6.0 mm cylindrical specimens for compression-compression fatigue measurements. The two end sides of each alloy were polished using a 1200 SiC grit-surface finish perpendicular to the longitudinal axis of the specimens using a polishing fixture to ensure that the sides were parallel to one another. A computer-controlled Material Test System (MTS) servohydraulic-testing machine was employed for fatigue studies. The sample was tested with a load ratio, R, of 10 where $R = \sigma_{min}/\sigma_{max}$ (where σ_{min} and σ_{max} are the applied minimum and maximum stresses, respectively) under a load-control mode, using a sinusoidal waveform at a frequency of 10 Hz. The neutron powder diffraction measurements were carried out at the Los Alamos National Laboratory using the high intensity powder diffractometer (HIPD). Data were collected in a closed-cycle refrigeration system as a function of temperature. The structure function, $S(Q)$, obtained from the neutron powder diffraction data was Fourier transformed to obtain the pair density function (PDF) analysis. The inelastic data were collected using the Disk Chopper Spectrometer (DCS) at the NIST Center for Neutron Research (NCNR) as a function of temperature. Two incident energies were used, λ_i = 2.2 and 3.6 Å while data before and after fatigue were collected for the $Zr_{50}Cu_{40}Al_{10}$ composition only.

4. CONCLUSIONS

From the above results, the following conclusion can be drawn regarding the effects of mechanical fatigue on the atomic properties. A connection is present between the changes observed in the local structure and the dynamics in the solid state. The consequence of the increase of the first PDF peak height with cooling above the expected increase from thermal contraction is twofold: it signals a local restructuring due to microscopic changes as well as reduction of the VDOS. The local restructuring occurs because the lattice changes irreversibly, even after the load is removed. With repeated loading and unloading the glass eventually breaks, thus such lattice irreversibility is a precursory effect to failure. The decrease in the intensity of the vibrational spectrum and its shift towards the elastic channel follow the local restructuring. The local structure reorganization that we observe with modifications in the pair correlations amplitudes will have a direct effect on the normal modes and can explain the reduction of the amplitude of fluctuations with fatigue. It is the case that the system becomes brittle after fatigue and thus, brittleness is correlated to the suppression of the vibrational spectrum.

ACKNOWLEDGMENTS

The work at the University of Virginia is supported by the National Science Foundation, under contract DMR-0804801, at the University of Tennessee under contracts DMR-0909037, CMMI-0900271 and 1100080, and at NIST, DCS is supported under contract DMR-0944772. The Los Alamos National Laboratory is operated by Los Alamos National Security LLC under DOE contract DE-AC52-06NA25396.

CONFLICT OF INTEREST

The authors declare no conflict of interest.

REFERENCES

1. Debenedetti, P.G.; Stillinger, F.H. Supercooled liquids and the glass transition. Nature 2001, 410, 259–267.

2. Spaepen, F. A microscopic mechanism for steady state inhomogeneous flow in metallic glasses. Acta Mater. 1977, 25, 407–415.

3. Argon, A.S. Plastic deformation in metallic glasses. Acta Mater. 1979, 27, 47–58.

4. Masumoto, T.; Maddin, R. Structural stability and mechanical properties of amorphous metals. Mater. Sci. Eng. 1975, 19, 1–24.

5. Greer, A.L.; Ma, E. Bulk metallic glasses: At the cutting edge of metals research. MRS Bull. 2007, 32, 611–615.

6. Chen, M.; Inoue, A.; Zhang, W.; Sakurai, T. Extraordinary plasticity of ductile bulk metallic glasses.Phys. Rev. Lett. 2006, 96, 245502:1–245502:4.

7. Scopigno, T.; Ruocco, G.; Sette, F.; Monaco, G. Is the fragility of a liquid embedded in the properties of its glass? Science 2003, 302, 849–852.

8. Das, J.; Tang, M.B.; Kim, K.B.; Theissmann, R.; Baier, F.; Wang, W.H.; Eckert, J. "Work-hardenable" ductile bulk metallic glass. Phys. Rev. Lett. 2005, 94, 205501:1–205501:4.

9. Greer, A.L. Metallic glasses. Science 1995, 267, 1947–1953.

10. Freels, M.; Wang, G.Y.; Zhang, W.; Liaw, P.K.; Inoue, A. Cyclic compression behavior of a Cu-Zr-Al-Ag bulk metallic glass. Intermetallics 2011, 19, 1174–1183.

11. Wang, G.Y.; Liaw, P.K.; Jin, X.Q.; Yokoyama, Y.; Huang, E.W.; Jiang, F.; Keer, L.M.; Inoue, A. Fatigue initiation and propagation behavior in bulk metallic glasses under a bending load. J. Appl. Phys. 2010,108, 113512:1–113512:7.

12. Wang, G.Y.; Qiao, D.C.; Yokoyama, Y.; Freels, M.; Inoue, A.; Liaw, P.K. Effects of loading modes on the fatigue behavior of Zr-based bulk metallic glasses. J. Alloys Comp. 2009, 483, 143–145.

13. Buchenau, U.; Nucker, N.; Dianoux, A.J. Neutron scattering study of the low-frequency vibrations in vitreous Silica. Phys. Rev. Lett. 1984, 53, 2316:1–2316:4.

14. Sette, F.; Krisch, M.H.; Masciovecchio, C.; Ruocco, G.; Monaco, G. Dynamics of glasses and glass-forming liquids studies by inelastic x-ray scattering. Science 1998, 280, 1550–1555.

15. Shintani, H.; Tanaka, H. Universal link between the boson peak and transverse phonons in glass.Nature Mat. 2008, 7, 870–877.

16. Sokolov, A.P.; Calemczuk, R.; Salce, B.; Kisliuk, A.; Quitmann, D.; Duval, E. Low temperature anomalies in strong and fragile glass formers. Phys. Rev. Lett. 1997, 78, 2405:1–2405:1.

17. Duval, E.; Boukenter, A.; Achibat, T. Vibrational dynamics and the structure of glasses. J. Phys. Condens. Matter. 1990, 2, 10227–10234.

18. Klinger, M.I.; Kosevich, A.M. Soft-mode-dynamics model of acoustic-like high-frequency excitations in boson-peak spectra of glasses. Phys. Lett. A 2001, 280, 365–370.

19. Chumakov, A.I.; Monaco, G.; Monaco, A.; Crichton, W.A.; Bosak, A.; Rüffer, R.; Meyer, A.; Kargl, F.; Comez, L.; Fioretto, D.; et al. Equivalence of the Boson peak in glasses to the transverse acoustic van hove singularity in crystals. Phys. Rev. Lett. 2011, 106, 225501:1–225501:4.

20. Schirmacher, W.; Diezemann, G.; Ganter, C. Harmonic vibrational excitations in disordered solids and the "Boson peak". Phys. Rev. Lett. 1998, 81, 136:1–136:4.

21. Taraskin, S.N.; Loh, Y.L.; Natarajan, G.; Elliott, S.R. Origin of the Boson peak in systems with lattice disorder. Phys. Rev. Lett. 2001, 86, 1255:1–1255:4.

22. Novikov, V.N.; Sokolov, A.P. Poisson's ratio and the fragility of glass-forming liquids. Nature 2004,431, 961–963.

23. Johnson, W.L. Bulk Glass-Forming Metallic Alloys: Science and Technology. In Bulk Metallic Glasses, Proceedings of Materials Research Society Symposium, Boston, MA, USA, December 1–3, 1998; Johnson, W.L., Inoue, A., Liu, C.T., Eds.; Materials Research Society: Warrendale, PA, USA, 1999; 554, pp. 191–196.

24. Inoue, A. Bulk Amorphous Alloys: Practical Characteristics and Applications; Trans Tech Publications: Zurich, Switzerland, 1999.

25. Pacard, C.E.; Homer, E.R.; Al-Aqeeli, N.; Schuh, C.A. Cyclic hardening of metallic glasses under Hertzian contacts: Experiments and STZ dynamics simulations. Philos. Mag. 2010, 90, 1373–1390.

26. Packard, C.E.; Witmer, L.M.; Schuh, C.A. Hardening of a metallic glass during cyclic loading in elastic range. Appl. Phys. Lett. 2008, 92, 171911:1–171911:3.

27. Li, Y.; Bai, H.Y.; Wang, W.H.; Samwer, K. Low-temperature specific heat anomalies associated with the boson peak in CuZr-based bulk metallic glasses. Phys. Rev. B 2006, 74, 052201:1–052201:4.

28. Inoue, A. Stabilization of metallic supercooled liquid and bulk amorphous alloys. Acta Mater. 2000, 48, 279–306.

29. Wang, G.Y.; Liaw, P.K.; Peter, W.H.; Yang, B.; Yokoyama, Y.; Benson, M.L.; Green, B.A.; Kirkham, M.J.; White, S.A.; Saleh, T.A.; et al. Fatigue behavior of bulk metallic glasses. Intermetallics 2004, 12, 885–892.

30. Qiao, D.C.; Huang, L.; Jiang, W.; Fan, C.; Liaw, P.K. Fatigue softening of (Zr58Ni13.6Cu18Al10.4)99Nb1bulk metallic glass. Metall. Mat. Trans. A 2010, 41, 1787–1791.

31. Tang, M.B.; Bai, H.Y.; Pan, M.X.; Zhao, D.Q.; Wang, W.H. Einstein oscillator in highly-random-packed bulk metallic glass. Appl. Phys. Lett. 2005, 86, 021910:1–021910:3.

32. Keppens, V.; Mandrus, D.; Sales, B.C.; Chakoumakos, B.C.; Dai, P.; Coldea, R.; Maple, M.B.; Gajewski, D.A.; Freeman, E.J.; Bennington, S. Localized vibrational modes in metallic solids. Nature 1998, 395, 876–878.

33. Hermann, R.P. Einstein oscillators in thallium filled antimony skutterudites. Phys. Rev. Lett. 2003, 90, 135505:1–135505:1.

34. Grigera, T.S.; Martin-Mayor, V.; Parisi, G.; Verrocchio, P. Phonon interpretation of the "Boson peak" in supercooled liquids. Nature 2003, 422, 289–292.

35. Malinovsky, V.K.; Sokolov, A.P. The nature of boson peak in Raman spectra in glasses. Solid State Comm. 1986, 57, 757–761.

36. Yannopoulos, S.N.; Andrikopoulos, K.S.; Ruocco, G. On the analysis of the vibrational Boson peak and low-energy excitations in glasses. J. Non-Cryst. Solids 2006, 352, 4541–4551.

37. Kazimirov, V.Y.; Louca, D.; Widom, M.; Gu, X.-J.; Poon, S.J.; Shiflet, G.J. Local organization and atomic clustering in multicomponent amorphous steels. Phys. Rev. B 2008, 78, 054112:1–054112:5.

38. Daw, M.S.; Baskes, M.I. Semiempirical, quantum mechanical calculation of hydrogen embrittlement in metals. Phys. Rev. Lett. 1983, 50, 1285:1–1285:4.

39. Cheng, Y.Q.; Ma, E.; Sheng, H.W. Atomic level structure in multicomponent bulk metallic glass. Phys. Rev. Lett. 2009, 102, 245501:1–245501:4.

9

PROBING INTERFACES IN METALS USING NEUTRON REFLECTOMETRY

Michael J. Demkowicz [1,2,*] and Jaroslaw Majewski [3,4]

[1]Department of Materials Science and Engineering, Massachusetts Institute of Technology, Cambridge, MA 02139, USA

[2]Materials Science and Engineering, Texas A & M University, College Station, TX 77843, USA

[3]MPA-CINT/Los Alamos Neutron Scattering Center, Los Alamos National Laboratory, Los Alamos, NM 87545, USA

[4]Department of Chemical Engineering, University of California at Davis, Davis, CA 95616, USA

ABSTRACT

Solid-state interfaces play a major role in a variety of material properties. They are especially important in determining the behavior of nano-structured materials, such as metallic multilayers. However, interface structure and properties remain poorly understood, in part because the experimental toolbox for characterizing them is limited. Neutron reflectometry (NR) offers unique opportunities for studying interfaces in metals due to the high penetration depth of neutrons and the non-monotonic dependence of their scattering cross-sections on atomic numbers. We review the basic physics of NR and outline the advantages that this method offers for investigating interface behavior in metals, especially under extreme environments. We then present several example NR studies to illustrate these advantages and discuss avenues for expanding the use of NR within the metals community.

Keywords: neutron reactor; spallation source; metals; extreme conditions

1. INTERFACES IN METALS

Metals form a wide variety of interfaces, including grain and phase boundaries [1], surface-liquid interfaces [2,3], solidification fronts [4], and mechanical contacts [5]. Although they typically occupy a small fraction of the total volume, interfaces play an outsized role in determining the properties of metals [6,7,8,9]. Understanding interfaces is therefore critical to predicting and controlling the behavior of metals.

Experimental investigation of interfaces presents significant challenges. Because they are often buried within the material, accessing them frequently requires destructive characterization or sample preparation methods, such as transmission electron microscopy (TEM) [10] or atom probe tomography (APT) [11]. Interfaces in metals typically have low thickness; indeed, some are atomically sharp [12]. Thus, characterizing them requires high—sometimes Å-level—spatial resolution. Moreover, certain interfaces only exist at high temperatures and pressures [13,14,15] or under contact with external media, such as gases or liquids [2,16]. Investigating such interfaces requires special *in situ* characterization methods.

An expanded experimental toolbox promises to accelerate progress in understanding metal interfaces, especially in extreme environments. This paper offers a primer on neutron reflectometry (NR): a characterization method with several advantages for studying metal interfaces [17,18,19,20,21]. NR is a mature experimental tool. The first NR experiments were conducted by Fermi and Zinn [22] and

Fermi and Marshall [23]. The technique has experienced continuous improvement since then [17,24,25,26]. Nevertheless, use of this method within the metals community has been relatively limited. We illustrate the potential benefits of NR to investigations of interfaces in metals by explaining the physics of the NR experiment and by presenting several example studies.

2. THE PHYSICS OF NEUTRON REFLECTOMETRY (NR)

Figure 1 shows a schematic of a typical NR measurement. The sample is a thin, planar film on a substrate. The experiment is usually conducted in air or vacuum, but may also be carried it out in other media (e.g., seeSection 6.4) [27]. The neutron source may be a fission nuclear reactor or a spallation source. In the nuclear reactor, the sustained nuclear fission of ^{235}U- or ^{239}Pu-rich fuels immersed in H_2O, D_2O, or solid graphite produces neutrons that may be used for scattering experiments. Spallation sources usually utilize pulsed high-energy (~GeV) protons to bombard targets made of heavy elements (such as W, Hg, U) to extract neutrons [28,29].

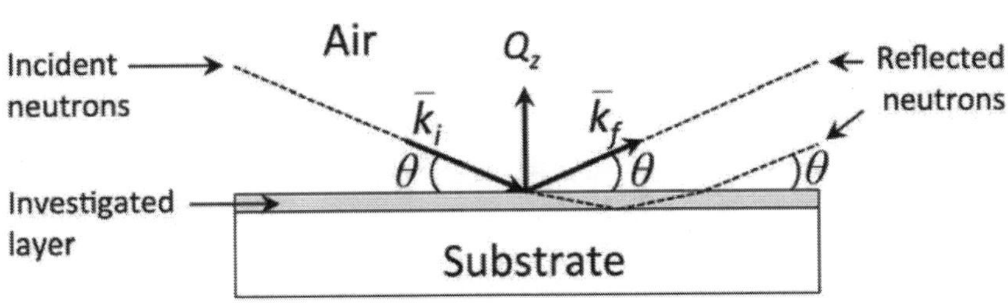

Figure 1. Schematic of a neutron reflectometry measurement.

In material property studies, the neutrons extracted from nuclear reactors or spallation sources are moderated to decrease their energies and therefore increase their wavelengths to Å ranges. Such moderation, depending on the final wavelength of neutrons required, is usually achieved by passing high-energy neutrons through H_2O, liquid H_2, or solid methane. Low energy neutrons are typically detected indirectly through absorption reactions with materials of high cross-sections for such reactions. Typically, ^3He, ^6Li, or ^{10}B is used to emit high-energy particles, whose ionization signatures may be detected by a number of means.

Upon interacting with the sample at a particular angle of incidence, θ (or a particular value of the neutron momentum transfer vector, Q_z), the incoming neutron beam can undergo absorption, reflection, transmission, or refraction. Consequently, there is a difference between the intensity of the outgoing, specularly reflected neutron beam and that of the incident beam. This difference—measured as a function of Q_z—encodes information about the distribution of the nuclear scattering length density along the direction normal to the sample surface. Moreover, the neutron is a ½-spin fermion and possesses a magnetic moment oppositely oriented to the spin. Therefore, its interaction with matter may depend on the sample's spin or magnetic field. Neutrons interact both with nuclear spins and the magnetic moments of unpaired electrons via dipole-dipole processes. Interactions with unpaired electrons may be of similar magnitude as nuclear scattering. However, they are not inherently isotropic. Rather, they depend on the orientation of the sample's magnetization *vis-a-vis* the direction of the neutron momentum wavevector transfer Q_z: only the component of sample's magnetization which is perpendicular to Q_z affects the neutron scattering. Therefore, the intensity of specularly reflected neutrons measured as a function of Q_z also encodes information about the distribution of the magnetization in the sample as a function of depth [30].

Depending on the specific NR technique, NR can take advantage of a range of different neutron-sample interactions [31]. However, this short review focuses on elastic specular NR, which is by far the most widely used NR technique. Elastic scattering conserves energy. Thus, we exclude any energy-dissipating neutron-matter interactions, except neutron absorption. In specular NR, the detector is positioned so as to measure outgoing neutrons at the same angle of incidence as the incoming neutrons, as illustrated in Figure 1. The experiment measures reflectivity, R, defined as the ratio of the number of reflected neutrons to the number of incoming neutrons.

The de Broglie expression $\lambda = \frac{h}{m_n v}$, relates the neutron's wavelength λ with its momentum, $p = m_n v$(where h is Planck's constant and m_n = 1.6749 × 10^{-27} kg is the neutron mass). Based on this formula, some simple relationships between the wavelength λ (Å), energy E (meV), and speed v (m/s) of the neutron can be developed: $E = 81.89/\lambda^2$ and $v = 3960/\lambda$. Thus, for example, a neutron with de Broglie wavelength of 1.5 Å has an energy E = 36.4 meV and velocity v = 2640 m/s. NR utilizes neutrons with wavelengths from sub-Å to tens of Å. By contrast, the size of an atomic nucleus is on the order of ~10 fm (1 fm = 10^{-15} m = 10^{-5} Å). Thus, to an excellent approximation, the incoming neutrons may be thought of as waves interacting with a uniform medium whose properties are determined by the density and type of atoms it contains. Their behavior may be described using Schrödinger's equation.

The assumption of elastic and specular conditions greatly simplifies the analysis the NR measurement. We further assume that the scattering properties of the sample vary only in one direction—namely, along the sample's normal—and therefore the components of the neutron wavevector parallel to the sample surface are not affected. Under these conditions, the component of the neutron wavevector parallel to the sample surface is conserved and the magnitude of the outgoing wavevector, $\vec{k}_f$, equals that of the incoming wavevector, $\vec{k}_i$: $\left|\vec{k}_i\right| = \left|\vec{k}_f\right| = \frac{2\pi}{\lambda}$. The difference between them, $\vec{Q}_z = \vec{k}_f - \vec{k}_i$ is known as the "momentum wavevector transfer" and lies perpendicular to the sample surface. From the geometry of the measurement (Figure 1), we calculate $\left|\vec{Q}_z\right| = Q_z = \frac{4\pi \sin(\theta)}{\lambda}$.

Quantum mechanics describes the incoming and outgoing neutron beams as a wavefunction, ψ, consisting of a superposition of plane waves:

$$\psi\left(\vec{x}\right) = e^{i\vec{k}_i \cdot \vec{x}} + re^{i\vec{k}_f \cdot \vec{x}} \quad (1)$$

Here, *r* is the amplitude of the outgoing (reflected) wave, normalized by the amplitude of the incoming wave (taken as unity). Knowing *r*, we may calculate reflectivity as R=|r|2. To compute *r*, however, we must model the interaction of the incoming neutron beam with the sample and substrate.

Because the component of the wavevector parallel to the sample surface is conserved, we may rewrite ψ as solely a function of *z*—the distance perpendicular to the sample surface—and k⊥i—the component of k⃗ i in the *z*-direction:

$$\psi\left(z\right) = e^{ik_i^{\perp}z} + re^{-ik_i^{\perp}z} \quad (2)$$

where we have used $k_f^{\perp} = -k_i^{\perp}$. Indeed, the entire NR measurement may be analyzed as a one-dimensional problem in the *z*-direction [32]. We write down wavefunctions of the type shown in Equation (2) for every distinct layer of material in the experiment, including air (or any other external medium), every layer of material in the sample, and the substrate (though in the substrate there is no reflected wave).

In free space, the neutron has kinetic energy $E_k = \frac{1}{2}m_n v^2$ and zero potential energy (if gravity is neglected). By contrast, within a material, it has a potential energy given by the Fermi pseudopotential [33]:

$$V_{\text{Fermi}} = \frac{h^2\beta}{2\pi m_n} \quad (3)$$

where β is the nuclear scattering length density (SLD). β describes the mean neutron scattering effectiveness of the material, which depends on the number of nuclei of type *i* per unit volume, N_i, and the coherent neutron scattering length, b_i:

$$\beta = \sum_i N_i b_i \tag{4}$$

b_i may be complex with its imaginary component describing absorption. Its real part may be either positive or negative, depending on the isotope [34,35]. Since neutrons have spin, their interactions with magnetic materials require an extended description that tracks changes in spin polarization [30]. The interaction of a neutron's spin with atomic magnetism (or other source of magnetic induction) can lead to magnetic scattering length density distributions that may be of the same order of magnitude as the nuclear scattering length densities. However, depending on the neutron's spin orientation (spin-up or spin-down)*vis-à-vis* the magnetic field of reference, the magnetic component is either added or subtracted from the nuclear one.

Often, samples investigated by NR may be described as a stack of discrete layers: each with its own composition, density, and thickness. In such cases, the NR experiment may be described with a 1D, time-independent Schrödinger equation on a piece-wise linear potential [19,26,32]. By matching wave functions and their derivatives at the interfaces between successive layers (as well as in the surrounding medium and in the substrate), one may solve for the amplitudes of all the waves in the setup. In particular, r—the amplitude of the outgoing wave measured at the sample surface—may be found. This amplitude comes about by coherent interference of partial waves on all interfaces in the film. From it, we determine the quantity measured by the neutron detector in Figure 1: reflectivity, R.

Thus, the SLD and thickness of the sample and substrate determine R. Since SLD in turn depends on composition and density, R is an indirect measure of these characteristics as well as the thickness of the individual material layers in the sample and substrate [17,26]. The goal of the reflectivity experiment is to measure $R(Q_z)$ and then infer $\beta(z)$ by fitting a model of the SLD distribution to the data. Q_z may be varied by changing the angle of incidence, θ (if the neutron beam is monochromatic, *i.e.*, λ = const.), or by changing the neutron wavelength, λ. The latter method of varying Q_z is typical of NR measurements at facilities where different neutron wavelengths, λ, are distinguished by the time of flight method. Figure 2 shows calculated reflectivity curves corresponding to a 500 Å Ni film on a quartz substrate in air. To illustrate the sensitivity of NR to isotopic composition, the calculation is carried out for two different isotopes: ^{58}Ni and ^{62}Ni.

Figure 2 illustrates some of the common features of reflectivity curves. Whenever the energy of the neutron is at or below the potential of the substrate (*i.e.*, whenever $k_i^2 \leqslant 4\pi\beta_{\text{substrate}}$), the neutrons are totally reflected from the surface. The onset of total reflection is called the critical edge and the value of Q_z at that point is referred to as Q_{critical}. The fringes in Figure 2 arise from interference between waves reflected from the top surface and the buried interface between the substrate and the layer. For this simple case, the spacing of the fringes may be calculated analytically: $d_{\text{fringe}} = \frac{2\pi}{t_{\text{layer}}}$, where t_{layer} is the thickness of the Ni layer. The amplitude of the fringes relates to the contrast between the layer and the substrate. The overall falloff of the curve obeys the Fresnel law: R~Qz−4. Most interfaces are not discontinuous, but rather graded due to chemical mixing or surface roughness. The surface roughness (which can be characterized by the root mean square displacement from the average interface, σ) may also be obtained from the reflectivity curve [36,37]. In general, the falloff of $R(Q_z)$ for rough or diffuse interfaces is even faster than that given by Fresnel's law.

3. INTERPRETING REFLECTIVITY CURVES

$R(Q_z)$ contains information about the SLD distribution perpendicular to the sample surface, β(z). Inferring β(z) from $\overline{R}(Q_z)$, however, is not trivial [38]. Because the NR measurement only collects the intensity of the reflected beam and not its phase, there is no unique mathematical transformation from $R(Q_z)$ to β(z). Therefore, NR data is usually interpreted by iteratively adjusting a trial SLD

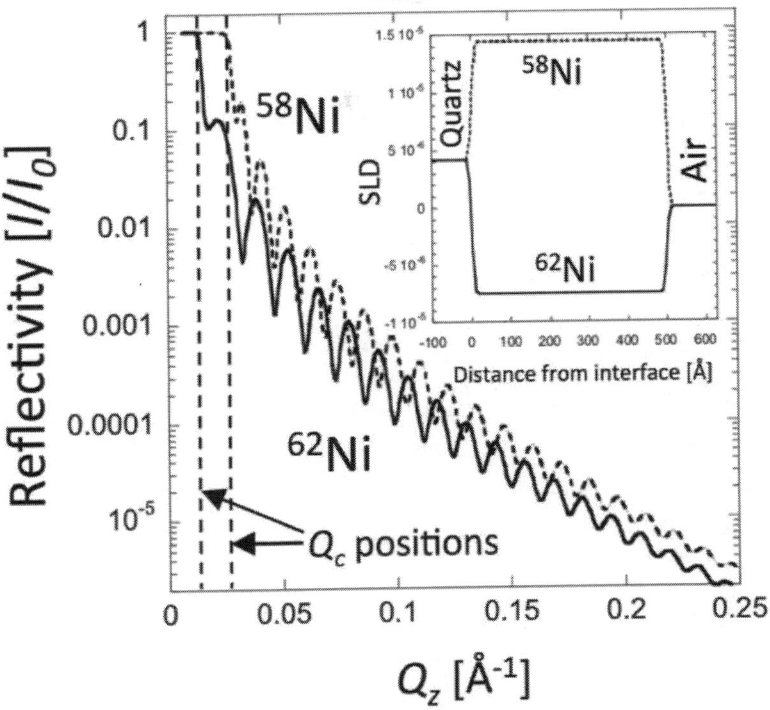

Figure 2. Calculated neutron reflectivity curves for 500 Å films of ^{58}Ni (dashed line) and ^{62}Ni (solid line) on a quartz substrate. The inset shows the SLD distributions for each isotope. The simulation was performed using MOTOFIT assuming RMS roughness parameters of 5 Å and the experimental resolution $\Delta Q_z/Q_z$ = 3%.

distribution, $\overline{\beta}(z)$, until the reflectivity it predicts, $\overline{R}(Q_z)$, matches the measured reflectivity, $R(Q_z)$, to within a specified tolerance.

The continuous function β(z) may often be approximated by a series of discrete layers—referred to as “boxes” or “slabs”—each with a constant SLD. Inter-layer roughness may be taken into account using an error function centered at each interface [36] or any other relevant functional form. A theoretical NR curve, $\overline{R}(Q_z)$, may be calculated from a trial SLD distribution, $\overline{\beta}(z)$, using the Parratt recursion formula [39,40], which relates the amplitudes of the reflected and transmitted waves at each interface. A number of approaches have been developed for adjusting $\overline{\beta}(z)$, to minimize the difference between $\overline{R}(Q_z)$, and $\overline{R}(Q_z)$, One example is the Levenburg-Marquardt nonlinear least-squares method used in open-source reflectivity package, MOTOFIT, which runs in the IGOR Pro environment [41]. This method seeks the least-squares fit of reflectivities, corresponding to a minimum χ^2 value. SLD models with the least number of boxes are usually preferred as they involve the smallest number of fitting parameters.

Once a best-fit set of model parameters is achieved, the uncertainties of these parameters may also be quantified by measuring the increase in χ^2 that comes about from perturbing each individual fitting parameter. For example, Reference [42] defines χ˜2 as the deviation of the reflectivity calculated using the perturbed parameter values from the best-fit reflectivity:

$$\tilde{\chi}^2 = \sum_{i=1}^{N} \left(\frac{y_i^{bf} - y_i^{p}}{y_i^{bf}} \right) \tag{5}$$

Here, ybfi is the best-fit to the measured reflectivity, ypi is the reflectivity value obtained by perturbing one parameter of the structural model, and *N* is the number of data points. The uncertainties on the fitting parameters are then defined as bounds within which χ˜2 is 5% or less.

Equation (4) shows that, at any given *z*, β(*z*) depends on the number and types of isotopes in the sample as well as their volume density. Thus, β(*z*) provides information concerning the composition and density at a depth *z*. Inferring these quantities from β(*z*) typically requires prior knowledge of some of the variables (e.g., nominal compositions or densities) or further input from other characterization methods. It should be re-emphasized that the SLD density profiles obtained from the fitting procedures described above are not unique. Due to the fact that only the intensities of scattered neutrons are measured in the NR experiment, but not their amplitudes and phases, there is no unique mathematical transformation leading from $R(Q_z)$ to SLD profile. Therefore, to resolve this problem, other data (e.g., from complementary characterization techniques) are often needed. In some cases, the phase of scattered neutrons may be resolved, as described by the work of Majkrzak and Berk [43,44,45] and others [46,47,48,49], enabling better inferences of SLD profiles.

4. ADVANTAGES OF NR

Neutron reflectometry offers unique advantages for characterizing solid-state interfaces in metals, including in extreme environments. Some of these advantages are easily deduced by considering the dependence of coherent scattering length, b_i, on atomic number, Z. Figure 3 shows that b_i is rather weakly dependent on Z. Indeed, the scattering lengths of almost all elements (in their natural isotopic abundance) are of the same order of magnitude. Therefore, in general, no one element can dominate the scattering of a multi-component sample, drowning out the contributions of other elements. In particular, light elements—such as H/D or He—may be detected, even when embedded in a matrix of heavy elements, e.g., of actinides. Moreover, NR is often able to distinguish elements with small differences in atomic number.

Figure 3 also shows that there are marked differences in b_i between different isotopes of certain elements. Thus, NR is well suited to investigations that require tracking of isotopes, such as tracer diffusion studies. Isotopic substitution may also aid in the interpretation of $\beta(z)$ profiles. For example, Ni has five different isotopes. The b_i for ^{58}Ni (abundance 68.3%) is 14.4×10^{-5} Å, whereas for ^{62}Ni (abundance 3.6%) b_i is negative and equals -8.7×10^{-5} Å. Substitution of ^{58}Ni (bulk SLD = 13.3×10^{-6} Å^{-2}) for ^{62}Ni (SLD = -7.5×10^{-6} Å^{-2}) significantly changes NR curves, as illustrated in Figure 2. Moreover, because neutron beams may be spin polarized, NR is especially well suited to investigations of magnetic properties of materials [50,51].

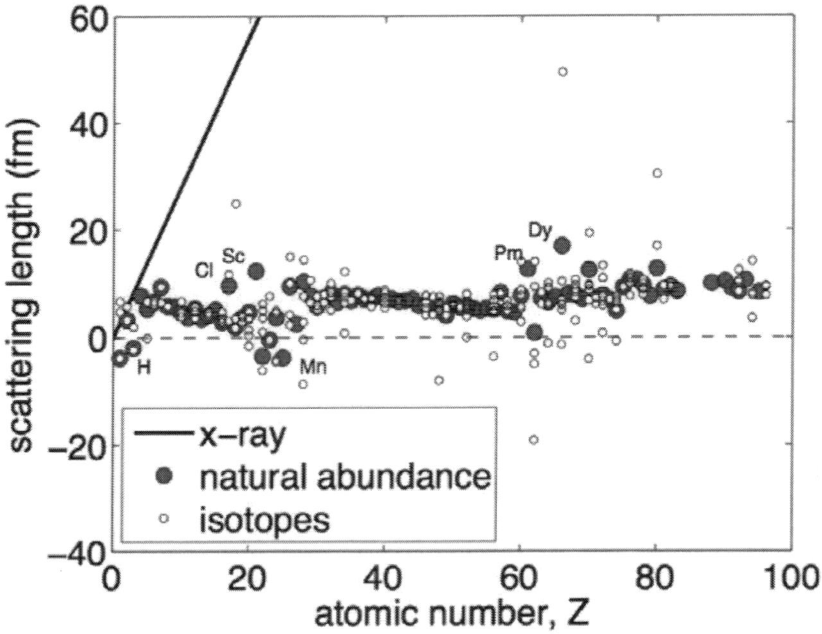

Figure 3. Neutron coherent scattering length, b_i, as a function of atomic number, Z. The X-ray scattering length is computed using Equation (6).

To appreciate the above-mentioned qualities, it is useful to compare NR with X-ray reflectometry (XRR) [52,53,54]. Although the mathematical description of XRR is similar to NR, the underlying physics of scattering is different: X-rays scatter from atoms' electrons while neutrons scatter off of atoms' nuclei. To interpret XRR experiments, we can replace the neutron coherent scattering lengths, b_i, in Equation (4) with the X-ray scattering lengths:

$$b_Z^{\text{X-ray}} = Z \cdot (2.81 fm) \tag{6}$$

where Z is the atomic number and 2.81 *fm* is the classical radius of the electron. Because X-rays scatter primarily from electrons, bX-rayZ is directly proportional to the number of electrons per atom, *i.e.*, to the atomic number, Z, in charge-neutral materials. Consequently, XRR cannot detect differences between isotopes. Moreover, because it is a linearly increasing function of *Z*, bX-rayZ provides very little contrast between elements with small *Z* differences. Finally, light elements—especially when embedded in high-*Z*matrices—are essentially undetectable with XRR, as the X-ray scattering length of the latter dominates reflectivity curves. However, it is often advantageous to use XRR and NR in tandem, as they may provide complementary information.

Equation (4) shows that scattering length density, β, is not only a function of b_i, but also of N_i: the number density of isotope *i*. Thus, depth profiles obtained by NR (and XRR) are sensitive not only to composition, but also to density. By contrast, depth profiles obtained through Rutherford backscattering are not sensitive to density [55]. NR is therefore capable of characterizing the evolution of porosity and of detecting displacive phase transformations that involve changes in density.

NR is also remarkable for its depth resolution, which is much greater than for XRR, especially for high-*Z*materials. Usually, NR techniques enable investigations of structures with total thickness up to ~3000 Å. For such thickness, the spacing between the scattering fringes (Figure 2) is very small, requiring very high $\Delta Q_z/Q_z$ resolution in the neutron detection. Typical $\Delta Q_z/Q_z$ values for existing neutron reflectometers are in the range from 2% to 5%. An instrumental resolution of $\Delta Q_z/Q_z = 2\%$ will result in the ability to distinguish between two thickness values which differ by 2%. Therefore differences of the film thickness on the order of Å can be readily detected. Smaller $\Delta Q_z/Q_z$ values may be achieved using detectors with higher spatial resolutions, better beam collimation, or better discrimination of the neutron wavelengths. However, increasing the resolution may result in smaller incident beam intensities, which can lead to longer measurements times and therefore higher scattering background. Thus, a proper balance between the two must be found. In general, it is advisable to adjust the resolution to match the expected thickness of the investigated films. Thick films, which give rise to dense oscillations of the interference fringes in $R(Q_z)$, require higher resolutions. By contrast, thin ones with broad interference oscillations in $R(Q_z)$ can be measured with lower resolution and therefore higher intensities, which can result in shorter measurement times and higher Q_z^{max} values.

Another advantage of NR is due to the ability of neutrons to penetrate deeply into solid matter. Several*mm* thick aluminum, quartz, silicon, or stainless steel windows absorb only a small fraction of incident neutrons with wavelengths in the Å range. Thus, NR measurements may be carried out to investigate the structure of buried interfaces as well as samples immersed in liquids or shielded from their environments by neutron-transparent containers. This quality is especially useful for investigating materials exposed to volatile media or under high pressure. Finally, NR frequently requires straightforward sample preparation and is not destructive. Therefore, samples investigated with NR may be subsequently further analyzed using other characterization methods. However, certain materials may be activated through interactions with the neutron beam, requiring some time for the radioactivity to decay before further characterization may be performed. It is also important to note that, since NR data are normalized to the incident neutron intensity, the measured SLD values are absolute.

5. PRACTICAL CONSIDERATIONS

Because neutron sources are inherently week (fluxes of ~10^{6-7} n/s/cm^2), the samples used for NR must be large. Samples with an area as small as 1 cm^2 may be measured, but at the expense of longer time of data acquisition and increased background noise. Neutron spallation sources usually provide some advantages enabling faster NR measurements. This is due to the polychromatic nature of the pulsed neutron beams they generate and the time of flight method used to discriminate between different neutron wavelengths, λ. NR spectra within a limited Q_z^{max} range (<0.1 Å^{-1}) can be obtained in 5–10 min. However, to obtain a full-spectrum NR data set (usually with $Q_z^{max} \approx 0.3$ Å^{-1} and $R \approx 10^{-6}$) requires from one to several hours of measurement time, regardless which neutron source is used.

Existing NR beamlines usually provide point-, line-, or 2-D neutron scattering detectors. Line- and 2-D detectors enable recording of scattering signals beyond the specular reflection: the so-called "off-specular" reflections. The "off-specular" data provides the neutron intensity distribution as function of the components of the neutron momentum transfer vector parallel to the sample's surface. This information can provide additional insight to extend the interpretation of the specular reflectivity measurements regarding in-plane correlations of the samples studied [56]. For example, these data allow correlations between the roughness of different interfaces or the growth of in-plane islands to be addressed. Reflectivities at high $Q_z$values are of great interest as they allow access to shorter length scales, which are important for characterizing the detailed structure of the investigated films. However, as already mentioned, the reflectivity R rapidly decays as Q_z^{-4}, making it difficult to acquire data at high Q_z. This challenge may be mitigated to some extent through the preparation of high quality samples: by minimizing the roughness of the sample surface as well as the roughness of its internal interfaces, high quality NR data at high Q_z and low R values may be collected. RMS roughness parameters up to 20 Å are usually tolerable, but detailed NR investigations typically require RMS roughness below 5 Å. Samples with such low roughness are most conveniently prepared using vapor deposition techniques. For such samples, $R \approx 10^{-6}$ and $Q_z^{max} \approx 0.2–0.3$ Å^{-1} can be routinely achieved for sample areas of several cm^2.

At the time of writing, there are several world-class NR instruments available worldwide, e.g., at the Spallation Neutron Source at Oak Ridge National Laboratory, the Lujan Center at Los Alamos National Laboratory, NCNR at NIST, the Institute Laue-Langevin in France, J-PARC in Japan, ANSTO in Australia, FRM-II in Germany, and several others. Several neutron sources are currently under construction or discussion. For example, the European Spallation Source in Sweden and the Second Target Station at SNS/ORNL will provide excellent capabilities for NR.

6. EXAMPLE APPLICATIONS OF NR TO METALS

This section provides examples of NR measurements conducted on metals. The examples are chosen to illustrate the unique advantages of the NR, namely its ability to detect density changes (Section 6.1), its sensitivity to magnetic moments and complementarity to X-ray reflectometry (XRR, Section 6.2), its sensitivity to light elements (Section 6.3), and its ability to penetrate through container walls (Section 6.4).

6.1. He in fcc/bcc Composites: Detecting Density Changes

Some nuclear transmutation reactions give rise to alpha particles, *i.e.*, nuclei of ^{4}He. When implanted into solids, these particles rapidly come to rest, pick up two electrons, and become regular He atoms. Since He is a noble gas and does not bond with surrounding atoms, it usually has negligible solubility within solids [57]. Thus, it precipitates out of solution into nanometer-scale bubbles [58]. These precipitates are usually deleterious to the properties of the solid, e.g., they lead to embrittlement in Ni-base alloys [59] and surface damage in plasma-facing materials [60]. Much effort has been invested into mitigating damage induced by implanted He, especially in materials for nuclear energy [61,62,63,64].

One way of controlling implanted He is to trap it at specially designed internal interfaces in composite materials [61]. However, investigations of this effect are limited by the difficulty of characterizing He precipitates at internal interfaces. NR (and XRR) provides a distinct advantage within this context: its sensitivity to local density changes enables detection of the onset of He precipitate formation [42,65,66].

Kashinath *et al.* investigated He precipitation at interfaces between copper (Cu) and one of three body-centered cubic (bcc) metals: niobium (Nb), vanadium (V), and molybdenum (Mo) [42]. They found that each of these interfaces has a distinct critical He dose at which precipitates begin to form. Figure 4 illustrates the findings of this study. Upper and lower bounds on best-fit SLD profiles were estimated by superimposing the upper and lower error bounds for each individual fitting parameter, as defined in Section 3. All SLD profiles with $\tilde{\chi}^2$ less than or equal to 5% are contained within these bounds, but the converse is not true: not all SLD profiles within these bounds have χ˜2 less than or equal to 5%. Therefore, these uncertainty estimates for best-fit SLD profiles are conservative.

The target is a Cu/Nb bilayer deposited on a Si substrate. Both the Cu and the Nb layer are approximately 20 nm thick. After implantation of 20 keV $^4He^+$ ions to a dose of 3 × $10^{16}/cm^2$, the reflectivity of the sample is consistent with an unaltered Cu/Nb bilayer structure, as shown in Figure 4a. However, upon implantation to a slightly higher He dose of 4 × $10^{16}/cm^2$, there is a clear change in the reflectivity, indicated by arrows inFigure 4a,b. This change may be explained by the formation of a layer of reduced density on the Cu side of the Cu–Nb interface, as illustrated in Figure 4b.

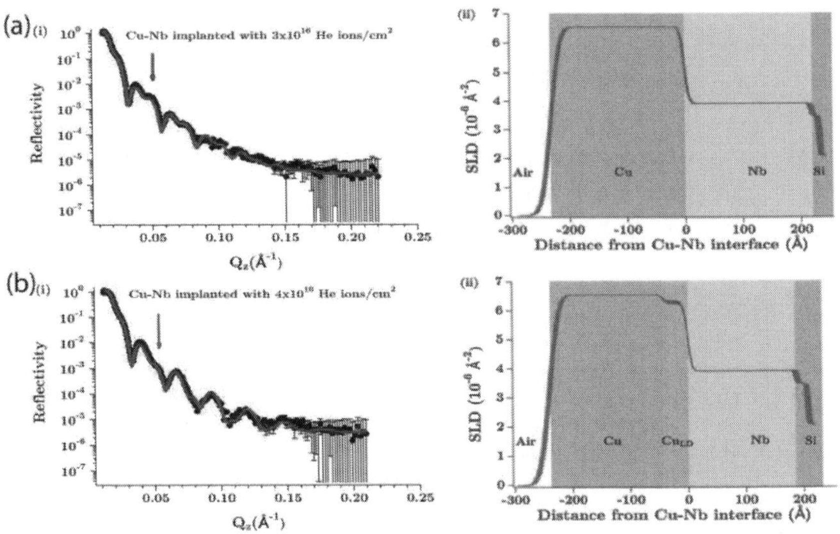

Figure 4. Reflectivity curves (**left** column) and SLD profiles (**right**column) for (**a**) 3 × $10^{16}/cm^2$ and (**b**) 3 × $10^{16}/cm^2$ He ions implanted in a Cu/Nb bilayer on a Si substrate. Reprinted with permission from Reference [42]. Copyright (2013), AIP publishing LLC.

At the He doses used in this study, nearly all the implanted He is believed to either escape through the Cu free surface or become trapped at the Cu–Nb or Nb–Si interfaces [42]. Precipitation within the Cu or Nb layers themselves is thought to be minimal. Thus, the low-density layer adjacent to the Cu–Nb interface inFigure 4b is thought to arise from the formation of He precipitates there. The critical He dose of $4 \times 10^{16}/cm^2$is consistent with preceding transmission electron microscopy (TEM) studies [61,67] as well as atomistic simulations [68]. However, whereas those previous investigations merely inferred interfacial precipitation, NR is able to observe it directly.

6.2. Fe/Y_2O_3 Interface: Sensitivity to Magnetization and Complementary to XRR

The structure of interfaces between low solubility metals—such as those discussed in the previous section—is easy to describe, as these interfaces are usually atomically sharp [12]. By contrast, the structure of oxide/oxide or metal/oxide interfaces is much more difficult to assess. Such interfaces are often several nanometers wide [69], exhibit transitions in structure reminiscent of phase changes [8], and contain intrinsic defects with distinct local compositions [70]. NR provides several advantages for investigating such interfaces, including high depth resolution and sensitivity to composition.

Watkins *et al.* used NR to study the structure of an interface between α-Fe and Y_2O_3 [71]. They found that this interface is a ~64 Å-thick transitional zone containing mixtures or compounds of Fe, Y, and O. By comparing their NR data to XRR and X-ray diffraction (XRD) measurements, they further determined that the interface was likely compositionally sharp upon synthesis and only later broadened as the neighboring crystals reacted. Finally, since α-Fe is ferromagnetic while Y_2O_3 is not, Watkins *et al.* were able to track changes in magnetization across this interface. Figure 5 shows that to model the reflectivity of this interface, contributions of spin-up and spin-down states of the neutron beam must be averaged. By using comparing the SLD profiles of these two states, the exact depth at which the ferromagnetic ordering is lost may be found (marked with an "x" in the right panel in Figure 5).

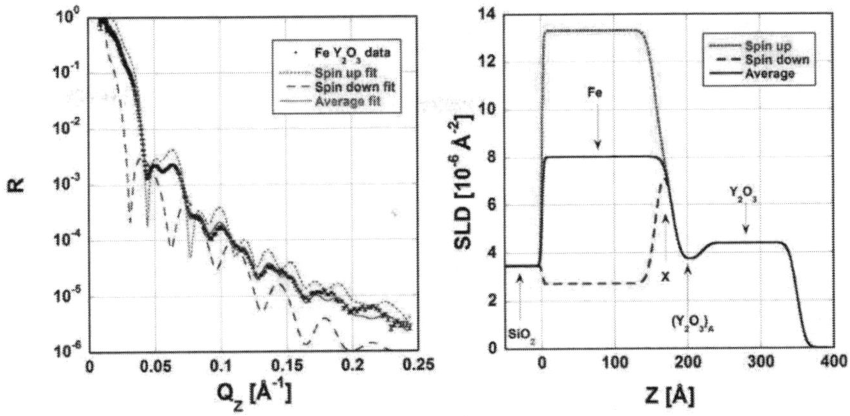

Figure 5. The neutron reflectivity (**left**) of the α-Fe and Y_2O_3 interface investigated by Watkins *et al.* is the average of SLD contributions from spin-up and spin-down states of the neutron beam (**right**). The depth at which ferromagnetic ordering is lost is marked with an "x" in the **right**panel. Also in the **right** panel, $(Y_2O_3)_A$ refers to a distinctive Y–O layer forming at this metal/oxide interface. Reprinted with permission from Reference [71]. Copyright (2014), AIP publishing LLC.

6.3. Actinides: Sensitivity to Light Elements

Actinides and their oxides exhibit some of the most intriguing and challenging chemistry known [72]. Frequently, the composition of these materials is not stoichiometrically precise. Moreover, their oxide structures can change dramatically under different environmental conditions. Neutrons provide a distinct advantage over X-rays in structural characterization of hydrides and oxides of heavy metals because they are better able to detect the lighter elements, such as H/D and O, within their actinide matrices. Figure 6illustrates neutron scattering length densities for different uranium oxide phases, showing that NR is able to distinguish between them.

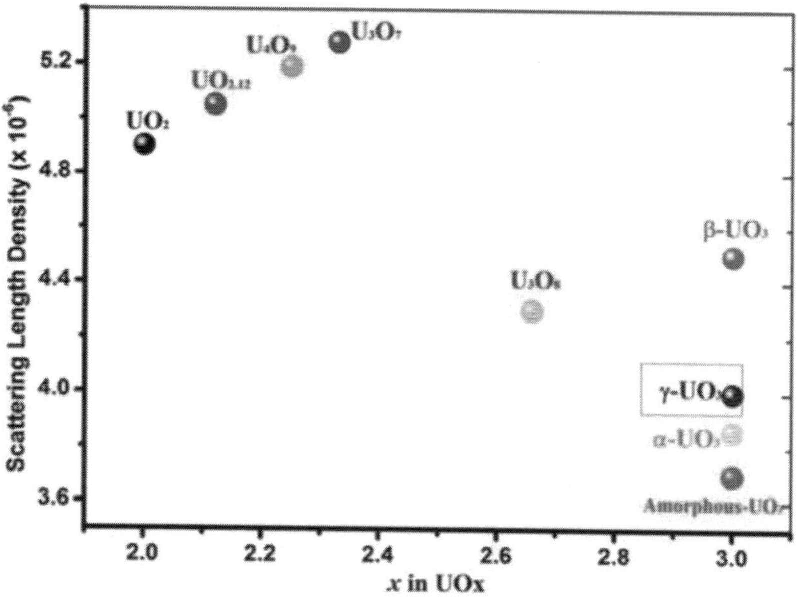

Figure 6. The calculated values of the nuclear scattering length density (SLD) for some common phases in uranium-oxide system. The calculations are based on specific densities published in literature [73,74].

The work of He *et al.* illustrates the utility of NR for investigating imperfect uranium oxide films [75]. They deposited uranium oxide on silicon substrates (with a thin native layer of oxide) using a combination of DC magnetron and reactive sputtering. In this technique, U atoms generated from a solid target by sputtering are readily oxidized by residual O_2 present in the Ar/O_2 mixture under moderate vacuum (approximately 1–3 × 10^{-4} torr) and then deposited on the substrate above the target. Several steps were taken to ensure high film quality: a multi-step sequential reactive deposition was used to minimize preferential film growth, the substrate was rotated to even out source distribution anomalies, and the partial pressures of Ar and O_2 were adjusted to control the composition of the uranium oxide. Nevertheless, the resulting film has a non-uniform, depth-dependent stoichiometry and structure.

The NR data for these films along with the best-fit curve according to the real-space SLD profile are shown in Figure 7a. According to these results, the total thickness of the UO_x film is about 630 Å. Figure 7b illustrates schematically the real-space structure represented by the best-fit SLD profile. The simplest model that fits the NR data has a three-layer structure. There is no heteroepitaxial growth of uranium oxide on the substrate/film interface due to the (~10 Å) native amorphous Si oxide layer on top of silicon wafer. The SLD of the layer at the film/air interface (~5.0×10^{-6} Å^{-2}) suggests the presence of hyper-stoichiometric phases. Meanwhile, the SLD of middle layer of the film (~3.8×10^{-6} Å^{-2}) together with the fact that no sharp X-ray diffraction peaks were observable (data not shown) indicates that this layer consists of amorphous α-UO_3. Overall, NR demonstrates a remarkably rich variation in structure and stoichiometry in this nominally uniform sample.

Another example of the utility of NR for studies of heavy metal (lanthanide) oxides arises from recent work on Dysprosium (Dy) oxidation [76]. They deposited Dy films on silicon substrates using the same DC magnetron sputtering technique as discussed above and characterized their structures using NR after exposure to air at two different temperatures: 25 °C (ambient temperature) and 150 °C. Figure 8 shows that, under both conditions, the film may be described three-slab model. Under ambient temperature, it consists of 20 Å silicon oxide on top of the Si substrate, 418 Å Dy, and 43 Å Dy_2O_3. After exposure to air in 150 °C for ~0.5 h, the thickness of the Dy_2O_3 increased to 114 Å while simultaneously the thickness of the Dy layer decreased to 363 Å. The total thickness of Dy and Dy_2O_3 layers increased from 461 Å to 477 Å, indicating an overall swelling of the sample. The roughness parameters of the air-Dy_2O_3 and Dy_2O_3-Dy interfaces decreased, making the top surface facing the air and the interface between the metallic Dy and its oxide smoother. For the two cases of uranium oxide and dysprosium studies described above, the approximate errors for the thickness, SLD, and roughness parameters vales were ±5 Å, 0.1 Å^{-2}, and 2 Å, respectively.

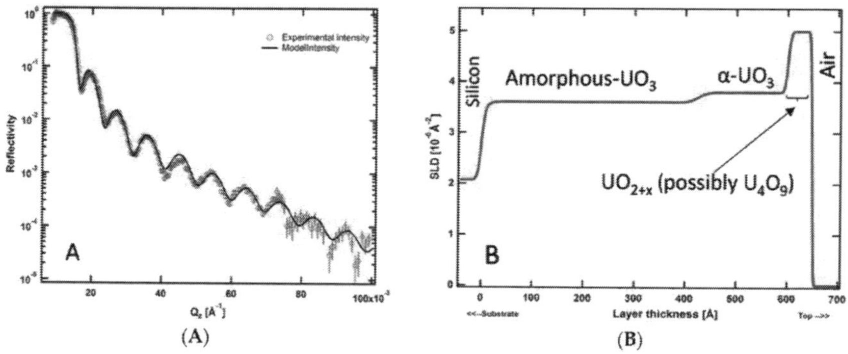

Figure 7. (**A**) NR data obtained from a UO_x deposited on a silicon substrate (open circles). Error bars indicate one standard deviation. The solid line through the data points corresponds to the best-fit SLD profile shown in (**B**).

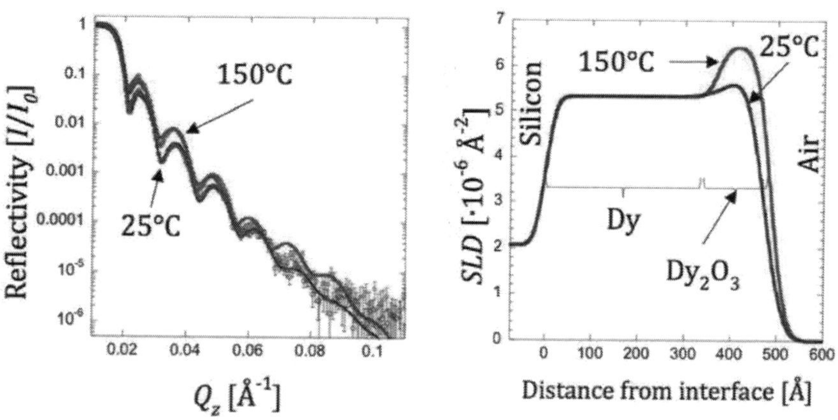

Figure 8. (**Left**) NR data (open circles) from a film of Dy covered with capping layer of Dy_2O_3 at ambient temperature (25 °C) and after expose to air at 150 °C for ~0.5 h. Error bars in the NR measurements correspond to one standard deviation. The solid line through the data points corresponds to the best-fit SLD profile shown on the (**right**).

6.4. Surfaces in Pressurized Liquids: Penetration of Neutrons through Containers

Neutrons penetrate through thick sections of solid matter with low attenuation. Thus, they are able to "see through" the walls of a high-pressure cell, enabling examination of metal surfaces in pressurized media [77]. Junghans *et al.* used this capability to study the corrosion of oxidized aluminum (Al) surfaces in pressurized seawater [27]. The corrosion of structural materials in the deep sea depends on numerous chemical and physical factors, including pH, dissolved oxygen, chloride ion activity, salinity, ocean currents, temperature, and hydrostatic pressure [78,79,80,81,82]. Al and its alloys find widespread use in marine environments, including in civil and defense vessels, offshore rigs, drill pipes for deep wells, and diving suits. There are several reports on corrosion of Al and its alloys at shallow depths, but investigations at high pressures are limited [78,79,80,82,83,84].

When in contact with air, Al develops a thin passivation layer of Al_2O_3. The surface oxide has a higher SLD than pure Al, providing neutron scattering contrast between the two materials (Al: $2.08 \cdot \times 10^{-6}$ Å^{-2}; crystalline Al_2O_3: $5.74 \cdot \times 10^{-6}$ Å^{-2}). Junghans *et al.* deposited uniform, ~900 Å-thick Al films on monocrystalline quartz wafers using DC magnetron sputtering and collected a total of 14 NR spectra over the course of 50 h (~3 h per spectrum). The film was in contact with 3.5 wt. % NaCl solution at pressures ranging from 1 to 600 atm in a specially developed solid/liquid, high pressure/temperature cell [77]. This cell provides the capabilities of solid/fluid interface investigations up to 2000 atm (~30,000 psi) and 200 °C. The cell's simple aluminum construction makes it easy to operate at high pressures and elevated temperatures, while the 13 mm thick neutron windows allow up to 74% neutron transmission. Figure 9 shows five representative NR measurements from this study [27].

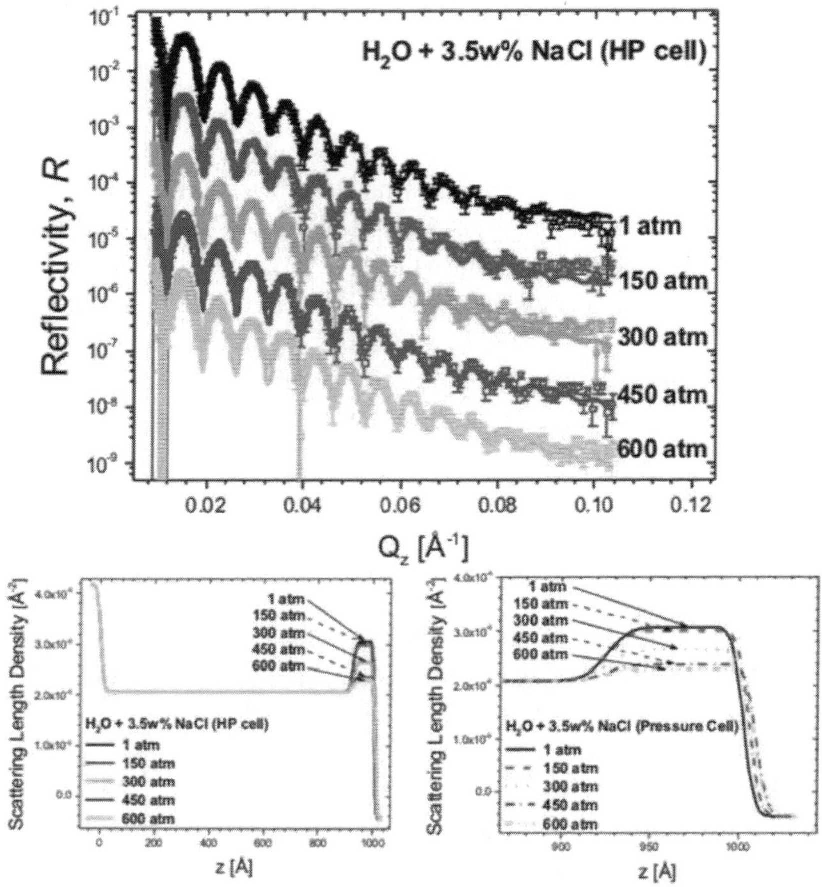

Figure 9. (**Top** panel) Five NR measurements of Al film deposited on quartz substrate and investigated at 25 °C in contact with H_2O + 3.5 wt. % of NaCl at different pressures. Solid curves are fits corresponding to the SLD profiles shown in the middle and right panels. Both the NR data and fits are offset by a decade along *y*-axis for clarity; (**Bottom left**panel) SLD distribution of the Al/Al_2O_3/liquid system and (**Bottomright** panel) magnified SLD distribution in the contact region. In both of the bottom panels, z = 0 at the quartz substrate/Al interface.

The NR results show virtually no corrosion of the Al layers. The observed decrease in the SLD of the starting Al_2O_3 passivation layer cannot be explained by the formation of new chemical compounds by the highly scattering Na, Cl, and O ions. However, this decrease is consistent with formation of stable Al–Cl–H_2O (or Al–O–Cl–H_2O) complexes or hydration of Al_2O_3 to $Al(OH)_3$. These results suggest that for the time scale of 50 h the influence of hydrostatic pressure only slightly influences interactions of the Al oxide film with Cl^- ions and H_2O. The corrosion rate is lower than reported by Beccaria *et al.* [78,79,80], suggesting slower kinetics for the reactions involved.

7. CONCLUSIONS

Neutron reflectometry (NR) is a mature experimental technique that has been used extensively in condensed-matter physics. However, its potential for investigating interfaces in metals has not yet been utilized widely. The present overview is intended to raise awareness of NR in the metals community with the hope of motivating wider use of this technique. NR provides several unique advantages for investigating interfaces in metals:

- **It is non-destructive**. Thus, NR results may be combined with other, follow-on investigations, e.g., using XRR, TEM, or APT.
- **Å-level depth resolution** enables detailed investigation of interface structure: thickness, SLD, and roughness of the layers.
- **Sensitivity to composition, isotopic distribution, density, and magnetic moment** allow multiple physical characteristics to be measured simultaneously.
- **Ability to detect low-*Z* elements**, such as H/D, He, and other light isotopes.
- **Suitability for *in situ* studies** due to the high penetrability of neutrons through container walls and surrounding media. This capability enables investigations of a variety of buried interfaces, including solid-liquid ones, which are otherwise very difficult to access with X-rays.
- **The measured SLD values are absolute** due to the fact that the reflected beam is normalized by the incident intensity of the neutron beam.

- **Ready access** thanks to the availability of several suitable neutron sources (reactors and spallation facilities) worldwide.

Several example applications of NR to metal surfaces and buried interfaces have been discussed above. As the metals community continues to explore the structure and properties of interfaces in ever-greater detail [7,8,9], NR stands poised to contribute valuable new additions to this ever-growing list of examples.

ACKNOWLEDGMENTS

J.M. thanks Heming He, Kirk Rector, Peng Wang, Ann Junghans, Erik Watkins (LANL), and David Allred (BYU). M.J.D. acknowledges Abishek Kashinath and support from the Center for Materials in Irradiation and Mechanical Extremes (CMIME), an Energy Frontier Research Center funded by the U.S. Department of Energy, Office of Science, Office of Basic Energy Sciences under Award No. 2008LANL1026. This work benefited from the use of the Lujan Neutron Scattering Center at LANSCE funded by the DOE Office of Basic Energy Sciences and Los Alamos National Laboratory under DOE Contract DE-AC52-06NA25396.

AUTHOR CONTRIBUTIONS

Both authors contributed equally to the composition of this article.

CONFLICTS OF INTEREST

The authors declare no conflict of interest.

REFERENCES

1. Sutton, A.P.; Balluffi, R.W. Interfaces in Crystalline Materials; Oxford University Press: Oxford, UK, 1995.

2. Furukawa, S.; de Feyter, S. Two-dimensional crystal engineering at the liquid-solid interface. InTemplates in Chemistry III; Broekmann, P., Dotz, K.H., Schalley, C.A., Eds.; Springer-Verlag Berlin: Berlin, Germany, 2009; Volume 287, pp. 87–133.

3. Kaplan, W.D.; Kauffmann, Y. Structural order in liquids induced by interfaces with crystals. Ann. Rev. Mater. Res. 2006, 36, 1–48.

4. Boettinger, W.J.; Warren, J.A.; Beckermann, C.; Karma, A. Phase-field simulation of solidification. Ann. Rev. Mater. Res. 2002, 32, 163–194.

5. Stachowiak, G.W.; Batchelor, A.W. Engineering Tribology; Butterworth-Heinemann: Boston, MA, USA, 2000.

6. Hirth, J.P.; Pond, R.C.; Hoagland, R.G.; Liu, X.Y.; Wang, J. Interface defects, reference spaces and the frank-bilby equation. Prog. Mater. Sci. 2013, 58, 749–823.

7. Beyerlein, I.J.; Demkowicz, M.J.; Misra, A.; Uberuaga, B.P. Defect-interface interactions. Prog. Mater. Sci. 2015, 74, 125–210.

8. Cantwell, P.R.; Tang, M.; Dillon, S.J.; Luo, J.; Rohrer, G.S.; Harmer, M.P. Grain boundary complexions.Acta Mater. 2014, 62, 1–48.

9. Mishin, Y.; Asta, M.; Li, J. Atomistic modeling of interfaces and their impact on microstructure and properties. Acta Mater. 2010, 58, 1117–1151.

10. Colliex, C.; Bocher, L.; de la Pena, F.; Gloter, A.; March, K.; Walls, M. Atomic-scale stem-eels mapping across functional interfaces. JOM 2010, 62, 53–57.

11. Schmitz, G.; Ene, C.; Galinski, H.; Schlesiger, R.; Stender, P. Nanoanalysis of interfacial chemistry.JOM 2010, 62, 58–63.

12. Mitchell, T.E.; Lu, Y.C.; Griffin, A.J.; Nastasi, M.; Kung, H. Structure and mechanical properties of copper/niobium multilayers. J. Am. Ceram. Soc. 1997, 80, 1673–1676.

13. Shaha, K.P.; Pei, Y.T.; Chen, C.Q.; Turkin, A.A.; Vainshtein, D.I.; de Hosson, J.T.M. On the dynamic roughening transition in nanocomposite film growth. Appl. Phys. Lett. 2009.

14. Olmsted, D.L.; Foiles, S.M.; Holm, E.A. Grain boundary interface roughening transition and its effect on grain boundary mobility for non-faceting boundaries. Scr. Mater. 2007, 57, 1161–1164.

15. Bellon, P. Nonequilibrium roughening and faceting of interfaces in driven alloys. Phys. Rev. Lett. 1998,81, 4176–4179.

16. Wiesler, D.G.; Majkrzak, C.F. Neutron reflectometry studies of surface oxidation. Phys. B 1994, 198, 181–186.

17. Russell, T.P. X-ray and neutron reflectivity for the investigation of polymers. Mater. Sci. Rep. 1990, 5, 171–271.

18. Dietrich, S.; Haase, A. Scattering of X-rays and neutrons at interfaces. Phys. Rep. Rev. Sec. Phys. Lett.1995, 260, 1–138.

19. Zhou, X.L.; Chen, S.H. Theoretical foundation of X-ray and neutron reflectometry. Phys. Rep. Rev. Sec. Phys. Lett. 1995, 257, 223–348.

20. Penfold, J.; Thomas, R.K. The application of the specular reflection of neutrons to the study of surfaces and interfaces. J. Phys. Condes. Matter 1990, 2, 1369–1412.

21. Van der Lee, A. Grazing incidence specular reflectivity: Theory, experiment, and applications. Solid State Sci. 2000, 2, 257–278.

22. Fermi, E.; Zinn, W.H. Reflection of neutrons on mirrors. Phys. Rev. 1946, 70, 103.

23. Fermi, E.; Marshall, L. Interference phenomena of slow neutrons. Phys. Rev. 1947, 71, 666–677.

24. Lekner, J. Theory of reflection of electromagnetic and particle waves. In Developments in Electromagnetic Theory and

Applications 3; Springer Netherlands: Dordrecht, The Netherlands, 1987.

25. Russell, T.P. The characterization of polymer interfaces. Annu. Rev. Mater. Sci. 1991, 21, 249–268.

26. Smith, G.S.; Majkrzak, C.F. Neutron reflectometry. In International Tables for Crystallography, 1st ed.; International Union of Crystallography, Ed.; Springer: Chester, UK; New York, NY, USA, 2006.

27. Junghans, A.; Chellappa, R.; Wang, P.; Majewski, J.; Luciano, G.; Marcelli, R.; Proietti, E. Neutron reflectometry studies of aluminum-saline water interface under hydrostatic pressure. Corros. Sci. 2015,90, 101–106.

28. Carpenter, J.M. Pulsed spallation neutron sources for slow-neutron scattering. Nucl. Instrum. Methods1977, 145, 91–113.

29. Bauer, G.S. Physics and technology of spallation neutron sources. Nucl. Instrum. Methods Phys. Res. Sect. A Accel. Spectrom. Dect. Assoc. Equip. 2001, 463, 505–543.

30. Hughes, D.J.; Burgy, M.T. Reflection of neutrons from magnetized mirrors. Phys. Rev. 1951, 81, 498–506.

31. Lauter, V.; Ambaye, H.; Goyette, R.; Lee, W.T.H.; Parizzi, A. Highlights from the magnetism reflectometer at the SNS. Phys. B 2009, 404, 2543–2546.

32. Merzbacher, E. Quantum Mechanics; Wiley: New York, NY, USA, 1961; p. 544.

33. Sears, V.F. Neutron scattering lengths and cross sections. Neutron News 1992, 3, 29–37.

34. Lovesey, S.W. Theory of Neutron Scattering from Condensed Matter; Clarendon Press: Oxfordshire, UK, 1984.

35. Neutron SLDs. Available online: http://www.ncnr.nist.gov/resources/n-lengths/ (accessed on 23 December 2015).

36. Nevot, L.; Croce, P. Characterization of surfaces by grazing X-ray reflection—Application to study of polishing of some silicate-glasses. Rev. Phys. Appl. 1980, 15, 761–779.

37. Als-Nielsen, J.; McMorrow, D. Elements of Modern X-ray Physics, 2nd ed.; Hoboken, N.J., Ed.; Wiley: Hoboken, NJ, USA, 2011.

38. Lovell, M.R.; Richardson, R.M. Analysis methods in neutron and X-ray reflectometry. Curr. Opin. Colloid Interface Sci. 1999, 4, 197–204.

39. Parratt, L.G. Surface studies of solids by total reflection of X-rays. Phys. Rev. 1954, 95, 359–369.

40. Yasaka, M. X-ray thin film measurement techniques. Rigaku J. 2010, 26, 1–9.

41. Nelson, A. Co-refinement of multiple-contrast neutron/X-ray reflectivity data using motofit. J. Appl. Crystallogr. 2006, 39, 273–276.

42. Kashinath, A.; Wang, P.; Majewski, J.; Baldwin, J.K.; Wang, Y.Q.; Demkowicz, M.J. Detection of helium bubble formation at fcc-bcc interfaces using neutron reflectometry. J. Appl. Phys. 2013.

43. Majkrzak, C.F.; Berk, N.F. Exact determination of the phase in neutron reflectometry. Phys. Rev. B1995, 52, 10827–10830.

44. Berk, N.F.; Majkrzak, C.F. Inverting specular neutron reflectivity from symmetric, compactly supported potentials. In Proceedings of the International Symposium on Advance in Neutron Optics and Related Research Facilities, Kumatori, Osaka, Japan, 19–21 March 1996; p. 107.

45. Majkrzak, C.F.; Berk, N.F. Exact determination of the phase in neutron reflectometry by variation of the surrounding media. Phys. Rev. B 1998, 58, 15416–15418.

46. Dehaan, V.O.; Vanwell, A.A.; Adenwalla, S.; Felcher, G.P. Retrieval of phase information in neutron reflectometry. Phys. Rev. B 1995, 52, 10831–10833.

47. Kasper, J.; Leeb, H.; Lipperheide, R. Phase determination in spin-polarized neutron specular reflection. Phys. Rev. Lett. 1998, 80, 2614–2617.

48. Lipperheide, R.; Kasper, J.; Leeb, H. Surface profiles from polarization measurements in neutron reflectometry. Phys. B 1998, 248, 366–371.

49. Leeb, H.; Grotz, H.; Kasper, J.; Lipperheide, R. Complete determination of the reflection coefficient in neutron specular reflection by absorptive nonmagnetic media. Phys. Rev. B 2001.

50. Majkrzak, C.F. Neutron scattering studies of magnetic thin films and multilayers. Phys. B 1996, 221, 342–356.

51. Williams, W.G. Polarized Neutrons; Clarendon Press: Oxford, UK; New York, NY, USA; Oxford University Press: Oxford, UK; New York, NY, USA, 1988.

52. Renaud, G.; Lazzari, R.; Leroy, F. Probing surface and interface morphology with grazing incidence small angle X-ray scattering. Surf. Sci. Rep. 2009, 64, 255–380.

53. Sinha, S.K. Reflectivity using neutrons or X-rays—A critical comparison. Phys. B 1991, 173, 25–34.

54. Stoev, K.N.; Sakurai, K. Review on grazing incidence X-ray spectrometry and reflectometry. Spectroc. Acta B Atom. Spectr. 1999, 54, 41–82.

55. Feldman, L.C.; Mayer, J.W. Fundamentals of Surface and Thin Film Analysis; North-Holland: New York, NY, USA, 1986.

56. Jablin, M.S.; Zhernenkov, M.; Toperverg, B.P.; Dubey, M.; Smith, H.L.; Vidyasagar, A.; Toomey, R.; Hurd, A.J.; Majewski, J. In-plane correlations in a polymer-supported lipid membrane measured by off-specular neutron scattering. Phys. Rev. Lett. 2011.

57. Laakmann, J.; Jung, P.; Uelhoff, W. Solubility of helium in gold. Acta Metall. 1987, 35, 2063–2069.

58. Trinkaus, H.; Singh, B.N. Helium accumulation in metals during irradiation—Where do we stand? J. Nucl. Mater. 2003, 323, 229–242.

59. Judge, C.D.; Gauquelin, N.; Walters, L.; Wright, M.; Cole, J.I.; Madden, J.; Botton, G.A.; Griffiths, M. Intergranular fracture in irradiated inconel X-750 containing very high concentrations of helium and hydrogen. J. Nucl. Mater. 2015, 457, 165–172.

60. Baldwin, M.J.; Doerner, R.P. Helium induced nanoscopic morphology on tungsten under fusion relevant plasma conditions. Nucl. Fusion 2008.

61. Demkowicz, M.J.; Misra, A.; Caro, A. The role of interface structure in controlling high helium concentrations. Curr. Opin. Solid State Mat. Sci. 2012, 16, 101–108.

62. Misra, A.; Demkowicz, M.J.; Zhang, X.; Hoagland, R.G. The radiation damage tolerance of ultra-high strength nanolayered composites. JOM 2007, 59, 62–65.

63. Odette, G.R.; Miao, P.; Edwards, D.J.; Yamamoto, T.; Kurtz, R.J.; Tanigawa, H. Helium transport, fate and management in nanostructured ferritic alloys: In situ helium implanter studies. J. Nucl. Mater.2011, 417, 1001–1004.

64. Odette, G.R.; Hoelzer, D.T. Irradiation-tolerant nanostructured ferritic alloys: Transforming helium from a liability to an asset. JOM 2010, 62, 84–92.

65. Zhernenkov, M.; Gill, S.; Stanic, V.; DiMasi, E.; Kisslinger, K.; Baldwin, J.K.; Misra, A.; Demkowicz, M.J.; Ecker, L. Design of radiation resistant metallic multilayers for advanced nuclear systems. Appl. Phys. Lett. 2014.

66. Zhernenkov, M.; Jablin, M.S.; Misra, A.; Nastasi, M.; Wang, Y.-Q.; Demkowicz, M.J.; Baldwin, J.K.; Majewski, J. Trapping of implanted he at Cu/Nb interfaces measured by neutron reflectometry. Appl. Phys. Lett. 2011.

67. Demkowicz, M.J.; Bhattacharyya, D.; Usov, I.; Wang, Y.Q.; Nastasi, M.; Misra, A. The effect of excess atomic volume on he bubble formation at fcc-bcc interfaces. Appl. Phys. Lett. 2010.

68. Kashinath, A.; Misra, A.; Demkowicz, M.J. Stable storage of helium in nanoscale platelets at semicoherent interfaces. Phys. Rev. Lett. 2013.

69. Kaplan, W.D.; Chatain, D.; Wynblatt, P.; Carter, W.C. A review of wetting versus adsorption, complexions, and related phenomena: The rosetta stone of wetting. J. Mater. Sci. 2013, 48, 5681–5717.

70. Dholabhai, P.P.; Pilania, G.; Aguiar, J.A.; Misra, A.; Uberuaga, B.P. Termination chemistry-driven dislocation structure at SrTiO3/MgO heterointerfaces. Nat. Commun. 2014.

71. Watkins, E.B.; Kashinath, A.; Wang, P.; Baldwin, J.K.; Majewski, J.; Demkowicz, M.J. Characterization of a Fe/Y2O3 metal/oxide interface using neutron and X-ray scattering. Appl. Phys. Lett. 2014.

72. Morss, L.R.; Edelstein, N.M.; Fuger, J.; Katz, J.J. The Chemistry of the Actinide and Transactinide Elements, 3rd ed.; Springer: Dordrecht, The Netherlands, 2006.

73. Hoekstra, H.R.; Siegel, S. The uranium-oxygen system—U3O8–UO3. J. Inorg. Nucl. Chem. 1961, 18, 154–165.

74. Loopstra, B.O.; Cordfunk, E. On structure of α-UO3. Recl. Trav. Chim. PaysBas 1966, 85, 135–142.

75. He, H.M.; Wang, P.; Allred, D.D.; Majewski, J.; Wilkerson, M.P.; Rector, K.D. Characterization of chemical speciation in ultrathin uranium oxide layered films. Anal. Chem. 2012, 84, 10380–10387.

76. Watkins, E.B.; Scott, B.; Allred, D.D.; Majewski, J. Unpublished work. 2015.

77. Wang, P.; Lerner, A.H.; Taylor, M.; Baldwin, J.K.; Grubbs, R.K.; Majewski, J.; Hickmott, D.D. High-pressure and high-

temperature neutron reflectometry cell for solid-fluid interface studies. Eur. Phys. J. Plus 2012.

78. Beccaria, A.M.; Poggi, G. Influence of hydrostatic-pressure on pitting of aluminum in sea-water. Br. Corros. J. 1985, 20, 183–186.

79. Beccaria, A.M.; Poggi, G. Effect of some surface treatments on kinetics of aluminum corrosion in NaCl solutions at various hydrostatic pressures. Br. Corros. J. 1986, 21, 19–22.

80. Beccaria, A.M.; Fiordiponti, P.; Mattogno, G. The effect of hydrostatic-pressure on the corrosion of nickel in slightly alkaline-solutions containing C1– ions. Corros. Sci. 1989, 29, 403–413.

81. Heusler, K.E. Untersuchungen der korrosion von aluminium in wasser bei hohen temperaturen und drucken. Mater. Corros. 1967, 18, 11–15.

82. Dexter, S.C. Effect of variations in sea-water upon the corrosion of aluminum. Corrosion 1980, 36, 423–432.

83. Venkatesan, R.; Venkatasamy, M.A.; Bhaskaran, T.A.; Dwarakadasa, E.S.; Ravindran, M. Corrosion of ferrous alloys in deep sea environments. Br. Corros. J. 2002, 37, 257–266.

84. Sawant, S.S.; Wagh, A.B. Corrosion behaviour of metals and alloys the waters of the arabian sea.Corros. Prev. Control 1990, 37, 154–157.

10

MONITORING OF BAINITE TRANSFORMATION USING *IN SITU* NEUTRON SCATTERING

Hikari Nishijima[1,†], Yo Tomota[2,*,‡], Yuhua Su[3], Wu Gong[3] and Jun-ichi Suzuki[4]

[1]Department of Applied Beam Science, Graduate school of Science and Enginieering, Ibaraki University, 4-12-1 Naka-narusawa, Hitachi 316-8511, Japan

[2]Graduate school of Science and Engineering, Ibaraki University, 4-12-1 Naka-narusawa, Hitachi 316-8511, Japan

[3]Japan Atomic Energy Agency, 2-4 Shirane Shirakata Tokai, Ibaraki 319-1195, Japan

[4]Comprehensive Research Organization for Science and Society, 162-1 Shirakata, Tokai, Ibaraki 319-1106, Japan

ABSTRACT

Bainite transformation behavior was monitored using simultaneous measurements of dilatometry and small angle neutron scattering (SANS). The volume fraction of bainitic ferrite was estimated from the SANS intensity, showing good agreement with the results of the dilatometry measurements. We propose a more advanced monitoring technique combining dilatometry, SANS and neutron diffraction.

Keywords: small angle neutron scattering; dilatometry; bainite transformation; *in situ* measurement; neutron diffraction

1. INTRODUCTION

Multi-phase steels containing carbon-enriched austenite have been extensively studied because of their excellent combination of strength and ductility/toughness. To obtain multi-phase structures, various material processings have been developed such as intercritical annealing followed by quenching to produce dual-phase steels [1,2,3], isothermal holding to yield carbon-enriched retained austenite for transformation-induced plasticity (TRIP steels) [4,5], isothermal holding at a low temperature to realize ultra-fine lamellar structure (nano-bainite steel) [6,7,8,9,10], and the partial quenching followed by up-heated isothermal holding (Q&P steels) [11,12,13,14,15]. In particular, nano-bainite steels consisting of nano-scale lamellae of bainitic ferrite and carbon-enriched austenite have attractively exhibited tensile strength greater than 2 GPa and fracture toughness of approximately 30 MPa $m^{1/2}$ [6,7]. The nano-bainite formed by isothermal holding at 300~400 °C shows an extremely slow transformation rate [6,16,17]. This heat treatment is favorable for producing large mechanical components with small residual stresses. However, the acceleration of the transformation must enlarge the application of nano-bainite steels. We have found that a small amount of low temperature ausforming (e.g., at 300 °C) [18,19] or partial quenching below Ms temperature [20] is effective to accelerate nano-bainite transformation. The dislocation structure introduced in austenite at low temperatures is found to assist bainite transformation with strong variant selection where partial dislocations introduced by ausforming play an important role for bainite transformation [19]. In these advanced steels, the processing is complicated enough that it is very important to monitor microstructure evolution quantitatively.

For studying microstructure evolution, *in situ* observations during processing using synchrotron X-ray [21,22] or neutron diffraction (ND) [19,20] have successfully been employed so far. Diffraction profiles provide the insights on volume fractions of constituents, which show good agreement with the results obtained by dilatometry [17], carbon contents in austenite [23], texture [19], and dislocation density [20]. Line broadening analysis for the ND profile provides "coherently diffracting mosaic size" probably related to dislocation cell size in engineering steels. The sizes larger than 1.0 μm such as austenite grain size or ferrite lath size cannot be evaluated by diffraction but hopefully by SANS or Bragg edge (BE) measurements.*In situ* BE measurement during bainite transformation was examined by Huang *et al.* [24]. They reported the changes in austenite volume fraction and carbon concentration with progress of bainite transformation but gave us no data on ferrite lath size. The two populations of austenite with different carbon contents have also been demonstrated by diffraction [19,22], but no information concerning the size of bainite lath has been reported so far. If we employ *in situ* SANS, the insights on the shape and size of the bainite lath would be obtained. In this study, we introduce a dilatometer into SANS-J-II at JRR-3/JAEA. The volume fraction of nano-bainite estimated from SANS data is compared with the results of conventional dilatometry. Then, we measure dilatometry, SANS and ND simultaneously to understand the mechanism of microstructure evolution during heat processing using an industrial neuron diffractometer, iMATERIA, at MLF/J-PARC. The traditional dilatometry provides only the phase fraction estimated from the amount of expansion or contraction of a specimen, whereas neutron experiments provide details in crystallography, chemical compositions, internal stresses, *etc.* This paper reports trials of such a monitoring system combined with complementary multi-methods.

2. EXPERIMENTAL PROCEDURES

2.1. Specimen Preparation

The chemical compositions of the steel used in this study were 0.79C–1.98Mn–1.51Si–0.98Cr–0.24Mo–1.06Al–1.58Co–balanced Fe (mass %). The steel was prepared by vacuum induction melting [8,9]. The ingot was homogenized at 1200 °C for 14.4 ks, followed by hot-rolling in the temperature range 1200–1000 °C to reduce the thickness from 40 mm to 10 mm through 10 successive rolling passes. Plate specimens with 15 × 15 × 1 mm^3 were prepared for SANS measurements at SANS-J-II (JRR-3/Japan Atomic Energy Agency, Tokai, Japan) and iMATERIA (Japan Proton Accelerator Research Complex (J-PARC), Tokai, Japan).

2.2. Small Angle Neutron Scattering Methods

In situ SANS measurements were performed using the SANS-J-II small angle neutron scattering instrument installed at the cold neutron beam line in the JRR-3 research reactor of the Japan Atomic Energy Agency (Tokai, Japan). For the SANS measurement, two two-dimensional (2D) detectors were used to detect neutrons scattered in the 0.005 to 0.199 nm^{-1} scattering vector q-range ($q = (4\pi/\lambda)\sin\theta$, where a half of scattering angle θ, and neutron wavelength $\lambda = 0.656$ nm), covering a real microstructure size of 3 to 1000 nm. The detector was positioned 10 m away from the specimen to measure the SANS profiles in the q-range of 0.005 to 0.237 nm^{-1}. Experimental set up is shown in Figure 1. As seen, a dilatometer and a 1.0 T magnet were installed for temperature control and separation of nuclear and magnetic scattering, respectively. A thermo-couple was spot-welded on the specimen surface to control temperature of the specimen. The specimen was heated up to 900 °C with a heating speed of 2 °C/s, held there to obtain an austenite single phase microstructure for starting, and then cooled down to 300 °C with a cooling rate of 5 °C/s, followed by isothermal holding at 300 °C in vacuum under a magnetic field of 1.0 T. The time interval for data acquisition was set to be 10 min (600 s) during the isothermal holding. Using the data in the parallel direction or vertical with respect to the magnetic field direction summing azimuthal sector within 30 degrees, SANS profile, *i.e.*, scattering intensity, was counted *versus* q to obtain profiles of nuclear and magnetic scattering components (for more details see Figure 3).

Concerning the iMATERIA at MLF/J-PARC, by which simultaneous measurements of dilatometry, ND and SANS were examined, the detailed explanation was omitted here because of a preliminary experiment. Brief explanation will be given in Section 3.3 together with some experimental results.

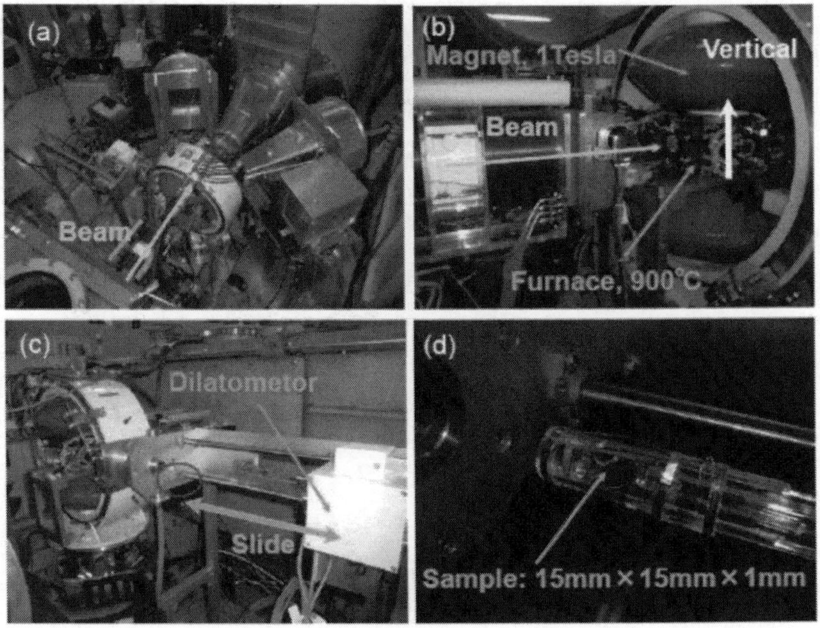

Figure 1. Experiment view of SANS with a 1 T magnet and a dilatometer to monitor the kinetics of bainite transformation at SANS-J-II/JRR-3 (JAEA): (**a**) overall top view; (**b**) magnet; (**c**) dilatometer and (**d**) specimen holder.

2.3. Data Analysis on Small Angle Neutron Scattering

The SANS intensities (I) obtained were plotted as a function of q, where lower q-values correspond to larger size of grain or particle. From the $\ln I(q)$ *versus* $\ln q$ plots, several microstructural parameters can be determined at different q regions, *i.e.*, the radius of gyration (R_g) representing "the effective size of the scattering particle" can be determined by the Guinier plot in the lower q region [25,26,27,28,29].

The q-range for the Guinier approximation depends on the particle size. The particle shape can be recognized from the q-dependence of the scattering intensity, *i.e.*, a slope of "−1" indicates cylinder shape, "−2" disc, and "−4" sphere. On the other hand, the Porod law holds in the high-q region and can be used to calculate other structural parameters. A slope of "−4" suggests that the interface of the particle is smooth and thereby the scattering intensity is proportional to the total interface area.

3. RESULTS AND DISCUSSION

3.1. Monitoring of Bainite Transformation by Dilatometry

Figure 2 shows the temperature history of a specimen measured with a thermo-couple and the change in length obtained by dilatometry (DL). As is usually performed, the data obtained by the traditional dilatometry indicate, apparently, dilatation caused by bainite transformation at 300 °C. Though the change in specimen length can be converted to the ferrite volume fraction, *i.e.*, kinetics of bainite transformation, the insights on chemical, crystallographic and microstructural features cannot be found. Therefore, as was described above, *in situ* neutron scattering/diffraction has been performed; *in situ* ND gave us the information on not only the change in the ferrite volume fraction but also the formation of two populations of austenite (the higher carbon concentration region and the lower one) [19,23], texture evolution [19], and dislocation density [20], as was mentioned above. Here, the size of the transformed product is expected to be monitored by SANS measurement.

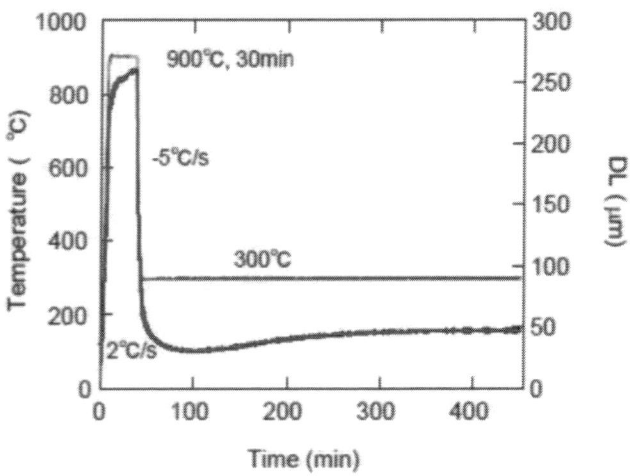

Figure 2. Change in temperature and the length of a specimen (DL) during heat treatment.

3.2. Monitoring of Bainite Transformation with in Situ SANS

Two-dimensional SANS patterns at different holding times are presented in Figure 3. It is found that the scattering intensity increases with the progress of bainite transformation both in the nuclear and magnetic components. These intensities were collected within 30° (see Figure 3c) in the parallel direction or vertical with respect to the magnetic field direction indicated in Figure 3a.

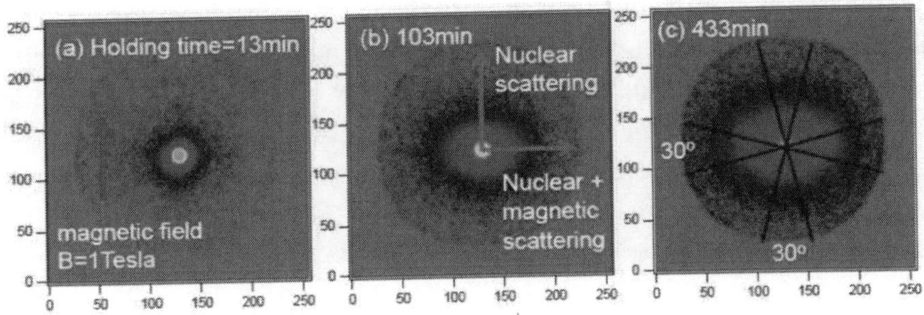

Figure 3. Change in the two-dimensional scattering pattern with holding time at 300 °C: (**a**) 13 min, (**b**) 103 min and (**c**) 433 min.

The nuclear component of the SANS profile obtained with a time interval of 10 min was plotted in Figure 4as a function of the *q*-value. After the onset of bainite transformation, the SANS intensity of the nuclear component increased, apparently, with the holding time, *i.e.*, progress of bainite transformation. Here, two interesting features are noticed; one is an increase in scattering intensity in a high *q* region, the so-called "Porod region" with a slope of −4, and the other is the slope change in a low *q* region, the so-called "Guinier region". In the Porod region, the scattering intensity is proportional to the total area of the interface of scattering inhomogeneity, the bainitic ferrite lath in the present case, so that the intensity increase means the progress of transformation.

The scattering intensities at lines A and B in Figure 4 were plotted as a function of holding time in Figure 5a,b, respectively. The curves are similar to the dilatation curve shown in Figure 2 and the change in volume fraction determined by *in situ* neutron diffraction in the previous study [18] (the results by dilatometry, SANS and ND obtained by iMATERIA are presented together in Figure 8). As is presented in Figure 6 as an example, the morphology of bainitic ferrite lath would be assumed by a disc shape with a different radius but the same thickness. Hence, the total area of the ferrite/austenite interface is postulated to be proportional to the ferrite volume fraction. This indicates that the scattering intensity in the Porod region is proportional to the ferrite volume fraction showing the bainite transformation kinetics. Because the ferrite phase is magnetic while the austenite phase is non-magnetic at 300 °C, the magnetic scattering component also increases with the increase of the ferrite volume fraction (compare curves labeled N and N + M), suggesting that the ferrite volume fraction could be determined not only by the nuclear scattering component but also by the magnetic scattering component. This means that a magnet is not needed for the evaluation of bainite transformation.

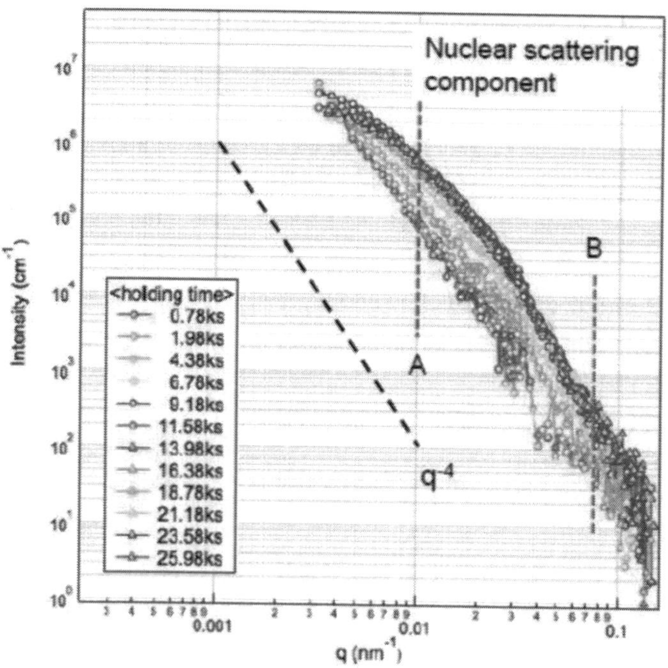

Figure 4. Change in nuclear component of SANS profile with progress in bainite transformation.

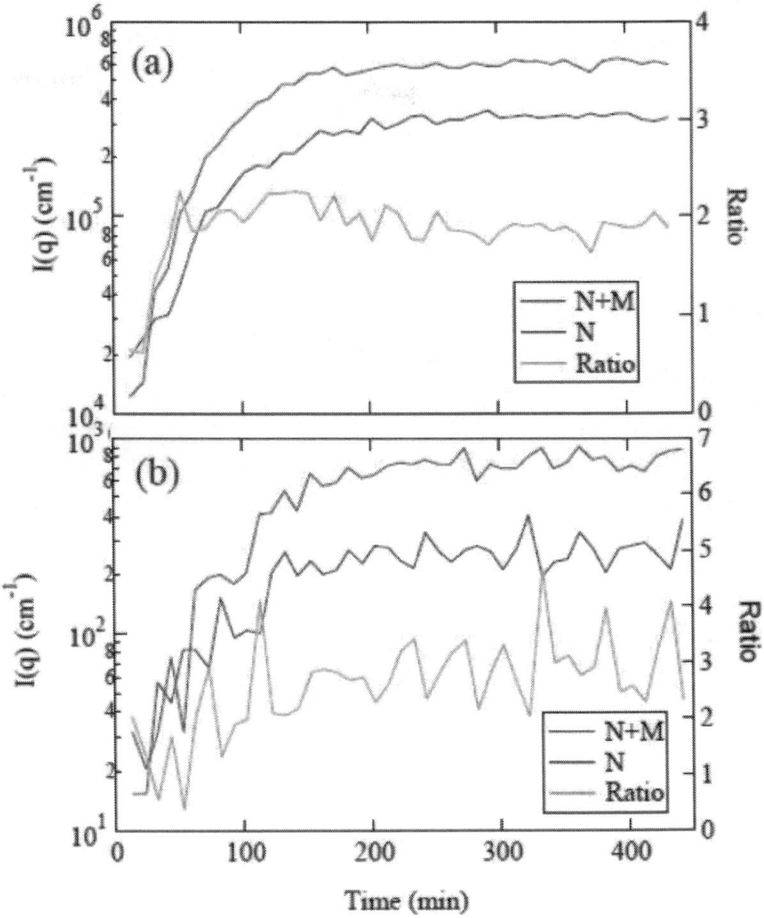

Figure 5. SANS intensity at the Porod region as a function of holding time at 300 °C: (**a**) at q = 0.01 nm^{-1} (line A in Figure 4) and (**b**) q = 0.073 nm^{-1} (line B in Figure 4).

In the Guinier region, the slope of the I–q curve would be −2 (disc shape) if it appeared completely. However, the size of the bainite lath is too large to be detected in the present q-range. Hence, as was done previously for non-metallic inclusions in steels [30], we need to expand the measuring q-range to cover smaller values. The influence of multiple scattering [31] is also suspected and, hence, we cannot follow the change in the size of the bainitic lath in this experiment.

From microstructure observations, it is very likely that the first lath is large in a scale of prior austenite grain size and that the later-formed laths must be shortened because the pre-formed laths inhibit further growth, although the thickness is nearly constant. Therefore, it is believed that the SANS intensity at the Porod region is proportional to the bainite volume fraction.

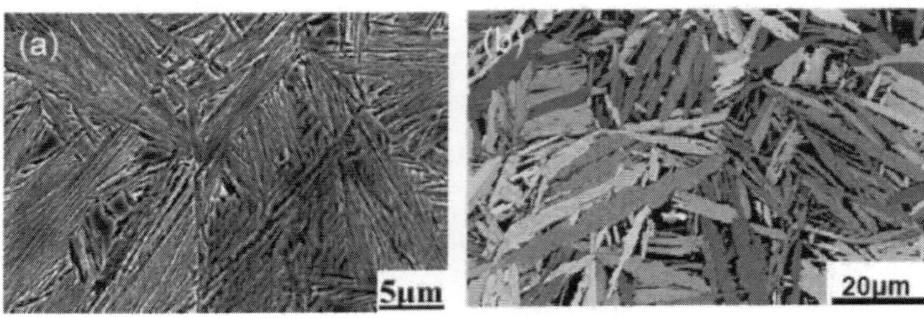

Figure 6. Microstructure of bainite formed at 300 °C: (**a**) SEM image and (**b**) EBSD/IPF map.

3.3. In Situ Measurements of Dilatometry, Small Angle Scattering and Diffraction at iMATERIA

In previous studies, we have successfully used *in situ* neutron diffraction to elucidate the transformation mechanism, particularly the effects of ausforming [18] and partial quenching [19]. The results in the previous Section 3.2 suggest that SANS is of use to monitor the transformation product. Hence, the combined measurements using the conventional dilatometry, ND and SANS were aimed at performing by introducing a new dilatometer into the engineering neutron diffractometer, iMATERIA at MLF/J-PARC, by which back-scatter ND and SANS can be measured simultaneously. A trial was performed using the same steel and some tentative results are presented here to show how to effectively do such an experiment.

Figure 7 shows the results of dilatometry and neutron diffraction obtained at the iMATERIA. As seen, the dilatometry result in Figure 7a is quite similar to that in Figure 2. Changes in austenite 111 and ferrite 110 diffraction profiles are presented in Figure 7b, in which the appearance of two populations of austenite with different amounts of carbon concentration found in the previous studies [19,20] is well confirmed because of the higher resolution of the back-scatter detector at the iMATERIA. The ferrite volume fraction (V_α) was calculated using Equation (1) [32,33] from the *hkl* diffraction intensities determined with the Z-Rietveld software (J-PARC, Tokai, Japan) [34].

$$V\alpha = \frac{\frac{1}{m}\sum^{m}\frac{I^{\alpha}_{hkl}}{R^{\alpha}_{hkl}}}{\frac{1}{m}\sum^{m}\frac{I^{\alpha}_{hkl}}{R^{\alpha}_{hkl}}+\frac{1}{n}\sum^{n}\frac{I^{\gamma}_{hkl}}{R^{\gamma}_{hkl}}} \quad (1)$$

where Iγhkl, Iαhkl, *n* and *m* refer to the measured integrated intensities of austenite, those of ferrite, the number of ferrite peaks and those of austenite, respectively, while Rαhkl and Rγhkl stand for theoretical values for texture-free material. The obtained results are plotted in Figure 7c showing a good coincidence with the dilatometry result in Figure 7a.

The data analysis for SANS measurements at the iMATERIA is now in progress, so that the scattering intensity at q = 0.4–0.42 nm^{-1} was tentatively counted. The scattering intensity obtained was plotted in Figure 8 as a function of holding time together with the results obtained by ND and dilatometry. As can be observed, these three results are in good agreement. This was the first trial to employ SANS for monitoring bainite transformation at the iMATERIA, and, hence, the detailed analysis has not been made yet; nuclear and magnetic components were not separated because a magnetic field was not applied to the specimen in this experiment.

Three methods of neutron scattering and diffraction are applicable to monitor bainite transformation, *i.e.*, ND, SANS and transmission BE [21] measurements. All of these methods can evaluate the transformation kinetics and have the following different possible features:

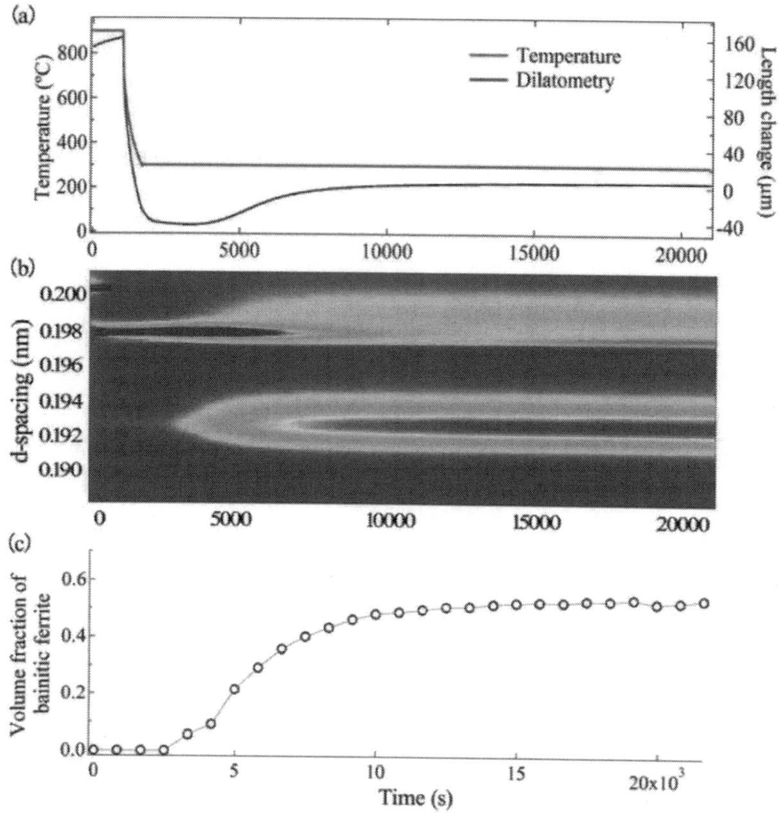

Figure 7. Results of preliminary *in situ* measurements during bainite transformation performed at iMATERIA: (**a**) temperature (red line) and dilatometric change (blue line); (**b**) changes in diffraction profile of austenite 111 and ferrite 110 peaks; and (**c**) volume fraction of ferrite determined from diffraction profiles.

(1)ND: crystal structure and volume fraction of constituents, texture, carbon concentration, elastic strain, and dislocation density: bulk average or three-dimensional (3D) distribution by scanning technique, although it requires scan time.

(2)SANS: volume fraction, shape and size of the second phase: bulk average.

(3)BE: phase volume fraction and carbon concentration of austenite, hopefully simultaneous 2D mapping of volume fraction, grain size, elastic strain and texture.

Hence, the combination of these methods would give us more fruitful information to understand microstructural evolution during processing for advanced steels.

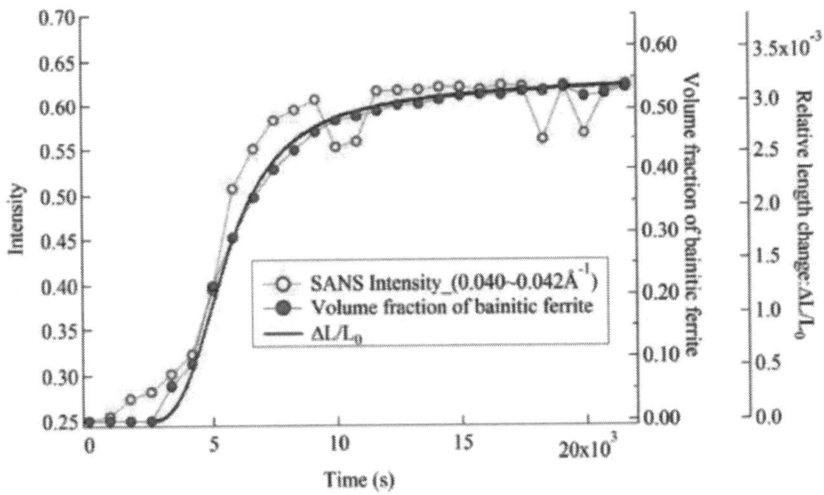

Figure 8. SANS intensity and volume fractions of ferrite determined by neutron diffraction and dilatometry as a function of holding time at 300 °C.

4. CONCLUSIONS

Bainite transformation was monitored by *in situ* measurements of conventional dilatometry as well as SANS and ND. The volume fraction of bainitic ferrite was estimated from the SANS intensity in the Porod region, showing good agreement with the results obtained by dilatometry and ND. A more advanced monitoring technique combining dilatometry, SANS, ND and BE was proposed.

ACKNOWLEDGMENTS

We would like to thank T. Ishigaki, S. Koizumi, A. Hoshikawa and K. Iwase of Ibaraki University (BL20 (iMATERIA) instrument scientists) for their help to install a dilatometer. The SANS experiments at JRR-3 was performed through the proposal #2010-A-72 and a preliminary experiment at MLF/J-PARC through proposal #2012PM0003. This study was financially supported by a Grant-in-Aid for Scientific Research A (# 21246106). The authors sincerely thank H. Beladi of Deaken University for supplying the present steel and valuable discussion during his stay at NIMS.

AUTHOR CONTRIBUTIONS

H.N. performed this research as her master degree thesis at Ibaraki University, Y.T. coordinated this research and wrote the manuscript with help from the other authors, Y. H. and W. G. helped with experiments and data analyses, and J.S. instructed the measurement method and analysis of SANS as an instrumental scientist.

CONFLICTS OF INTEREST

The authors declare no conflict of interest.

REFERENCES

1. Furukawa, T. Structure-property relationships of dual phase steels. J. Jpn. Inst. Met. 1980, 19, 439–446.
2. Tomota, Y.; Tamura, I. Strength and deformation behavior of two-ductile- phase alloys. J. Jpn. Inst. Met. 1985, 14, 657–664.
3. Tomota, Y.; Tamura, I. Mechanical Behavior of Steels Consisting of Two Ductile Phases. Trans. ISIJ1982, 22, 665–677.
4. Matsumura, O.; Sakuma, Y.; Takechi, H. Enhancement of elongation by retained austenite in intercritical annealed 0.4C–1.5Si–0.8Mn steel. Trans. ISIJ 1987, 27, 570–579.

5. Tomota, Y.; Tokuda, H.; Adachi, Y.; Wakita, M.; Minakawa, N.; Moriai, A.; Morii, Y. Tensile Behavior of TRIP-Aided Multi-Phase Steels Studied by in situ Neutron Diffraction. Acta Mater. 2004, 52, 5737–5745.

6. Caballero, F.G.; Bhadeshia, H.K.D.H.; Mawella, J.A.; Jones, D.G.; Brown, P. Very strong low temperature bainite. Mater. Sci. Technol. 2002, 18, 279–284.

7. Garcia-Mateo, C.; Caballero, F.G.; Bhadeshia, H.K.D.H. Development of Hard Bainite. ISIJ Int. 2003,43, 1238–1243.

8. Beladi, H.; Adachi, Y.; Timokhina, I.; Hodgson, P.D. Crystallographic analysis of nanobainitic steels.Scr. Mater. 2009, 60, 455–458.

9. Timokhina, I.; Beladi, H.; Xiong, X.Y.; Adachi, Y.; Hodgson, P.D. Nanoscale microstructural characterization of a nanobainitic steel. Acta Mater. 2011, 59, 5511–5522.

10. Carcia-Mateo, C.; Jimenz, J.A.; Yen, H.-W.; Miller, M.K.; Morales-Rivas, L.; Kuntz, M.; Ringer, S.P.; Yang, J.-R.; Caballero, F.G. Low temperature bainitic ferrite: Evidence of carbon super-saturation and tetragonality. Acta Mater. 2015, 91, 162–173.

11. Speer, J.G.; Streicher, A.M.; Matlock, D.K.; Rizzo, F.C.; Krauss, G. Austenite Formation and Decomposition; Damm, E.B., Merwinm, M., Eds.; TMS/ISS: Warrendale, PA, USA, 2003; pp. 502–522.

12. Speer, J.; Matlock, D.; de Cooman, B.; Schroth, J. Carbon partitioning into austenite after martensite transformation. Acta Mater. 2003, 51, 2611–2622.

13. Edmonds, D.; He, K.; Rizzo, F.; de Cooman, B.; Matlock, D.; Speer, J. Quenching and partitioning martensite—A novel steel heat treatment. Mater. Sci. Eng. A 2006, 438, 25–34.

14. Santofimia, M.J.; Zhao, L.; Sietsma, J. Overview of mechanism involved during the quenching and partitioning process in steels. Metall. Mater. Trans. A 2011, 42A, 3620–3626.

15. Yuan, L.; Ponge, D.; Wittig, J.; Choi, P.; Jimenez, J.A.; Raabe, D. Nanoscale austenite reversion through partitioning, segregation and kinetic freezing: Example of a ductile 2 GPa Fe–Cr–C steel. Acta Mater.2012, 60, 2790–2804.

16. Caballero, F.G.; Bhadeshia, H.K.D.H. Very Strong Bainite. Curr. Opin. Solid State Mater. Sci. 2004, 8, 251–257.

17. Koo, M.S.; Xu, P.; Suzuki, H.; Tomota, Y. Bainitic transformation behavior studied by simultaneous neutron diffraction and dilatometric measurement. Scr. Mater. 2009, 61, 797–800.

18. Gong, W.; Tomota, Y.; Koo, M.S.; Adachi, Y. Effect of ausforming on nanobainite steel. Scri. Mater.2010, 63, 819–822.

19. Gong, W.; Tomota, Y.; Adachi, Y.; Paradowska, A.M.; Kelleher, J.F.; Zhang, S.Y. Effects of ausforming temperature on bainite transformation, microstructure and variant selection in nanobainite steel. Acta Mater. 2013, 61, 4142–4154.

20. Gong, W.; Tomota, Y.; Harjo, S.; Su, Y.H.; Aizawa, K. Effect of prior martensite on bainite transformation in nanobainite steel. Acta Mater. 2015, 85, 243–249.

21. Babu, S.S.; Specht, E.D.; David, S.A.; Karapetrova, E.; Zachack, P.; Peer, M.; Bhadeshia, H.K.D.H. In situ Observations of Lattice Parameter Fluctuations in Austenite and Transformation to Bainite. Metall. Mater. Trans. A 2005, 36A, 3281–3289.

22. Stone, H.J.; Peet, M.J.; Bhadeshia, H.K.D.H.; Withers, P.J.; Babu, S.S.; Specht, E.D. Synchrotron X-ray studies of austenite and bainitic ferrite. Proc. Roy. Soc. A 2008, 464, 1009–1027.

23. Gong, W. Transformation Kinetics and Crystallography of Nano-bainite Steel. Ph.D. Thesis, Ibaraki University, Hitachi, Japan, March 2012.

24. Huang, J.; Vogel, S.C.; Poole, W.J.; Militzer, M.; Jacques, P. The study of low-temperature austenite decomposition in a Fe–C–Mn–Si steel using the neutron Bragg edge transmission technique. Acta Mater. 2007, 55, 2683–2691.

25. Muller, G.; Uhlemann, M.; Ulbricht, A.; Bohmert, J. Influence of hydrogen on the toughness of irradiated reactor pressure vessel steels. J. Nucl. Mater. 2006, 359, 114–121.

26. Ohnuma, M.; Suzuki, J.; Wei, F.G.; Tsuzaki, K. Direct observation of hydrogen trapped by NbC in steel using small-angle neutron scattering. Scr. Mater. 2008, 58, 142–145.

27. Buckley, C.E.; Birnbaum, H.K.; Bellmann, D.; Staron, P. Calculation of the radial distribution function of bubbles in the aluminum hydrogen system. J. Alloy. Compd. 1999, 293–295, 231–236.

28. Yasuhara, H.; Sato, K.; Toji, Y.; Ohnuma, M.; Suzuki, J.; Tomota, Y. Size Analysis of Nanometer Titanium Carbide in Steel by Using Small-Angle Neutron Scattering. Tetsu-to-Hagane 2010, 96, 545–549.

29. Su, Y.H.; Morooka, S.; Ohnuma, M.; Suzuki, J.; Tomota, Y. Quantitative Analyses on Cementite Spheroidization in Pearlite Structure by Small-Angle Neutron Scattering. Metall. Mater. Trans. A 2015,46A, 1731–1740.

30. Oba, Y.; Koppoju, S.; Ohnuma, M.; Kinjo, Y.; Morooka, S.; Tomota, Y.; Suzuki, J.; Yamaguchi, D.; Koizumi, S.; Sato, M.; et al. Quantitative Analysis of Inclusions in Low Carbon Free Cutting Steel Using Small-Angle X-ray and Neutron Scattering. ISIJ Int. 2012, 52, 458–464.

31. Schelten, J.; Schmatz, W. Multipe-Scattering Treatment for Small Angle Scattering Problem. J. Appl. Cryst. 1980, 13, 385–390.

32. Järvinen, M. Texture Effect in X-ray Analysis of Retained Austenite in Steels. Textures Microstruct. 1996,26–27, 93–101.

33. Gnäuperl-Herold, T.; Creuziger, A. Diffraction study of the retained austenite content in TRIP steels.Mater. Sc. Eng. A 2011, 528, 3594–3600.

34. Oishi, T.; Yonemura, M.; Morishima, T.; Oshikawa, A.; Torii, S.; Ishigaki, T.; Kamiyama, T. Application of matrix decomposition algorithms for singular matrices to Pawley method in Z-Rietveld. J. Appl. Cryst. 2012, 45, 299–308.

INDEX